Compendium of Pepper Diseases

Edited by

Ken Pernezny

Everglades Research and Education Center
University of Florida, Belle Glade

Pamela D. Roberts

Southwest Florida Research and Education Center
University of Florida, Immokalee

John F. Murphy

Auburn University, Auburn, Alabama

Natalie P. Goldberg

New Mexico State University, Las Cruces

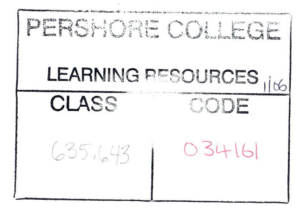

APS
PRESS

The American Phytopathological Society

Financial Sponsors

AgraQuest Inc.
Dow AgroSciences
DuPont Crop Protection
Marathon – Agricultural and Environmental Consulting, Inc.

Front cover photograph by Jeffrey R. Brushwein, used by permission

Back cover photographs by Ken Pernezny

Library of Congress Control Number: 2002115971
International Standard Book Number: 0-89054-300-3

Printed in the United States of America on acid-free paper

The American Phytopathological Society
3340 Pilot Knob Road
St. Paul, Minnesota 55121-2097, USA

Preface

This compendium presents diagnostic information on pepper diseases intended for plant pathologists and other professional agricultural scientists as well as county agents, scouts, crop consultants, commercial growers, and amateur horticulturists. A key feature of the book is the selection of color photographs depicting symptoms and signs of the important diseases of pepper worldwide.

The compendium is divided into five parts. Part I presents descriptions of diseases caused by infectious agents: bacteria, fungi, viruses, nematodes, and an angiosperm. Postharvest diseases and disorders are described in a separate section in Part I. Part II describes selected arthropod pests; Part III, abiotic and physiological disorders; Part IV, herbicide injuries; and Part V, nutritional disorders.

The descriptions of diseases and disorders include information about their distribution and importance, symptoms, causal agents, epidemiology, and control. Recommendations for disease and pest control, especially by teatment with agrochemicals, are of a general nature, since such information rapidly becomes outdated.

The authors of sections of the book are listed at the ends of the sections. The black-and-white and color illustrations are reproduced from photographs contributed by the authors of the sections in which the illustrations are mentioned, unless acknowledgment of another source is given.

The editors express sincere thanks to Tanya Simon and Janice Collins for their significant contributions to the preparation of the manuscript and to Jeff Jones, D. J. Jones, and Randy Ploetz for editorial support. The editors also acknowledge the facilities and time provided by the University of Florida, Auburn University, and New Mexico State University for work on this endeavor.

Authors

S. Adkins, U.S. Department of Agriculture, Fort Pierce, Florida

S. A. Alexander, Eastern Shore Agricultural Research and Education Center, Virginia Polytechnic Institute and State University, Painter

P. A. Banks, New Mexico State University, Las Cruces

J. A. Bartz, University of Florida, Gainesville

G. L. Benny, University of Florida, Gainesville

L. L. Black, Asian Vegetable Research and Development Center, Shanhua, Tainan, Taiwan

P. W. Bosland, New Mexico State University, Las Cruces

J. K. Brown, University of Arizona, Tucson

R. Creamer, New Mexico State University, Las Cruces

N. Goldberg, New Mexico State University, Las Cruces

S. K. Green, Asian Vegetable Research and Development Center, Shanhua, Tainan, Taiwan

P. T. Himmel, Seminis Vegetable Seeds, Woodland, California

G. J. Hochmuth, North Florida Research and Education Center, University of Florida, Quincy

J. B. Jones, University of Florida, Gainesville

T. A. Kucharek, University of Florida, Gainesville

E. M. Lamb, Indian River Research and Education Center, University of Florida, Fort Pierce

C. A. Lopes, Embrapa Hostalicas, Brasília, Brazil

M. Luis-Arteaga, Servicio de Investigación Agroalimentaria, Diputación General de Aragón, Zaragoza, Spain

D. N. Maynard, Gulf Coast Research and Education Center, University of Florida, Bradenton

R. McSorley, University of Florida, Gainesville

S. A. Miller, Ohio Agricultural Research and Development Center, Ohio State University, Wooster

M. T. Momol, North Florida Research and Education Center, University of Florida, Quincy

J. F. Murphy, Auburn University, Auburn, Alabama

G. Nuessly, Everglades Research and Education Center, University of Florida, Belle Glade

A. Obradović, Center for Vegetable Crops, Smad Palanka, Yugoslavia

K. Pernezny, Everglades Research and Education Center, University of Florida, Belle Glade

D. M. Persley, Queensland Horticulture Institute, Indooroopilly, Queensland, Australia

F. Ponz, Instituto Nacional de Investigación y Tecnología Agraria y Alimentaria, Madrid, Spain

P. M. Pradhanang, North Florida Research and Education Center, University of Florida, Quincy

B. B. Reddick, University of Tennessee, Knoxville

J. B. Ristaino, North Carolina State University, Raleigh

P. D. Roberts, Southwest Florida Research and Education Center, University of Florida, Immokalee

E. N. Rosskopf, U.S. Department of Agriculture, Fort Pierce, Florida

J. Schroeder, New Mexico State University, Las Cruces

S. H. Thomas, New Mexico State University, Las Cruces

R. A. Valverde, Louisiana State University, Baton Rouge

C. E. Warren, Auburn University, Auburn, Alabama

H. C. Wien, Cornell University, Ithaca, New York

T. A. Zitter, Cornell University, Ithaca, New York

Contents

Introduction
1 Significance of Pepper Worldwide
1 Origin
1 Botany and Culture

Part I. Diseases Caused by Infectious Agents
5 **Diseases Caused by Bacteria**
5 Bacterial Canker
6 Bacterial Spot
7 Bacterial Wilt
8 Syringae Seedling Blight and Leaf Spot
9 **Diseases Caused by Fungi and Oomycetes**
9 Anthracnose
10 Cercospora Leaf Spot (Frogeye Leaf Spot)
10 Charcoal Rot
11 Choanephora Blight (Wet Rot)
12 Damping-Off and Root Rot
13 Fusarium Stem Rot
14 Fusarium Wilt
15 Gray Leaf Spot
16 Gray Mold
17 Phytophthora Blight
19 Powdery Mildew
20 Southern Blight
21 Verticillium Wilt
22 White Mold
23 **Diseases Caused by Viruses**
24 *Alfalfa mosaic virus*
26 *Andean potato mottle virus* Pepper Strain
26 *Beet curly top virus*
27 Capsicum chlorosis virus
27 *Chilli veinal mottle virus*
28 *Chino del tomate virus*
29 *Cucumber mosaic virus*
31 *Pepper golden mosaic virus*
32 *Pepper huasteco yellow vein virus*
32 *Pepper mild mottle virus*
33 *Pepper mottle virus*
34 *Pepper veinal mottle virus*
35 *Potato virus Y*
36 *Sinaloa tomato leaf curl virus*
38 *Tobacco etch virus*
38 *Tobacco mosaic virus* and *Tomato mosaic virus*
39 *Tomato spotted wilt virus*
40 **Postharvest Diseases and Disorders**
41 Bacterial Soft Rot
42 Alternaria Rot
43 Botrytis Fruit Rot
44 Rhizopus Rot

45 Chilling Injury
45 **Disease Caused by an Angiosperm**
45 Dodder
46 **Diseases Caused by Nematodes**
46 Root-Knot Nematodes
47 Sting Nematode
48 Other Nematodes

Part II. Damage Caused by Arthropods
50 Aphids
50 Broad Mite
51 Thrips
51 True Bugs
52 Whiteflies

Part III. Abiotic and Physiological Disorders
53 Abnormal Fruit Shape
53 Blossom-End Rot
53 Color Spotting
54 Flower and Flower Bud Drop
54 Fruit Cracking
54 Hail Injury
54 Salt Injury
54 Sunscald
55 Wind Injury

Part IV. Herbicide Injury
56 Symptoms Caused by Herbicides Registered for Use on Peppers
56 Symptoms Caused by Accidental Application of Phytotoxic Herbicides
56 Symptoms Caused by Drift of Phytotoxic Herbicides
57 Symptoms Caused by Carryover of Residual Herbicides

Part V. Nutritional Disorders
59 Nitrogen
59 Potassium
59 Phosphorus
59 Calcium
59 Magnesium
60 Sulfur
60 Iron
60 Manganese

61 **Index**

Color plates (following page **30**)

v

Introduction

Whether they are called peppers, chiles, paprika, or ajis, plants in the genus *Capsicum,* like most crops, are afflicted with diseases, disorders, and pests that can reduce fruit quality and yield. Not all pepper disorders and pests occur in the same geographical area or at the same time. However, pests that reduce pepper yields are present in every production region. Pest management is one of the most important factors in producing an economical yield of peppers. In order to choose the proper treatment, it is important to diagnose the problem correctly. A wrong diagnosis may lead to the selection of a treatment that is inappropriate and expensive. Pests are best controlled by taking action before problems become serious. This book should help growers, extension agents, crop consultants, and researchers to correctly identify the causes of diseases and disorders of peppers.

Disease and insect control must start before pepper plants and seeds reach the field, and it is essential that a long-range pest management program be in place. Crop rotation is one of the best ways to promote healthy pepper production, since it helps minimize diseases, especially those caused by soilborne pathogens. Proper plant spacing to provide adequate movement of air around plants helps reduce the severity of foliar diseases. An equally important control method is planting disease-resistant pepper cultivars. Resistance is considered the most prudent means of disease control, because of its effectiveness, ease of use, and lack of potential negative effects on the environment. In addition, planting healthy seeds and transplants, ensuring a sufficient supply of water in the root zone and adequate drainage of the soil, controlling insects that vector disease agents, and maintaining sanitary conditions (for example, by cleaning and disinfecting equipment) will help to produce an economical crop of high-quality peppers.

Both nonparasitic and parasitic agents can cause pepper disease and injury. Parasitic pathogens causing disease include bacteria, fungi, phytoplasmas, viruses, parasitic higher plants, insects, nematodes, birds, and mammals. Nonparasitic factors causing disorders in peppers include extremes of temperature, moisture, and pH; high salt content of the soil; excessive or insufficient sunlight; lightning; deficient or excessive levels of nutrients; air pollutants; and pesticides.

Significance of Pepper Worldwide

Pepper fruit is consumed worldwide as a fresh vegetable or dehydrated for use as a spice. By volume, red pepper products, pungent and nonpungent, are one of the most important spice commodities in the world. The production of pepper, measured by tonnage, exceeds that of any other spice. Peppers add flavor and color to foods while providing essential vitamins and minerals. In many parts of the world, they provide the only variety to enhance diets that are otherwise bland. The range of food products that contain pepper or its chemical constituents is broad, including meats, salad dressings, mayonnaise, dairy products, beverages, candies, baked goods, snack foods, bread-ing and batters, salsas, and hot sauces. Pepper extracts are also used in pharmaceuticals and cosmetics. In addition to their uses as food, condiment, and medicine, peppers are grown as ornamentals in the garden.

Origin

There is agreement among taxonomists and historians that pepper originated in the tropical Americas. Archeological evidence indicates that it was one of the first crops domesticated in the Western Hemisphere. Chile seeds and chile fragments in coprolites dating back 9,000 years have been found in excavations in Tamaulipas and Tehuacán, Mexico. A handful of seeds of *Capsicum baccatum* found in a South American cave was estimated to be 7,000 years old. Peppers were carried to Spain by Columbus on his return trip in 1493, and Spanish and Portuguese traders were largely responsible for disseminating them to Africa and Asia.

Botany and Culture

Peppers belong to the family Solanaceae and the genus *Capsicum.* Currently, *Capsicum* includes 25 species, five of which have been domesticated. The most economically important and most widely cultivated species is *C. annuum* L. The center of genetic diversity of this species is in Mexico; a secondary center is in Mesoamerica. The other four domesticated species are *C. baccatum* L. and *C. pubescens* Ruiz & Pav., which were domesticated in the Peruvian Andes, and *C. chinense* Jacq. and *C. frutescens* L., which were domesticated in the Amazon Basin and in Central America, respectively. The other 20 species are used sparingly by humans but are a great reservoir of genes for resistance to diseases and pests.

Classification of peppers below the species level is based on fruit types, which are distinguished mostly by their characteristic shapes but also by their uses. Peppers can also be categorized by fruit color, pungency, aroma, and flavor. These subspecific categories are used in the pepper industry to aid in supplying the correct peppers for different products. Peppers of the diverse fruit types are very similar in their susceptibility to various disorders. However, breeders have introduced resistance into cultivars of some fruit types. For example, some bell pepper cultivars have been bred for resistance to bacterial leaf spot, viruses, and *Phytophthora.* For detailed information on pepper types and uses the reader is referred to *Peppers of the World* and *The Pepper Garden,* both by D. DeWitt and P. W. Bosland (see Selected References).

Climate
Peppers are herbaceous plants that can develop a shrublike habit. They are a shallow-rooted, warm-season crop requiring about the same growing conditions as eggplants and tomatoes.

Peppers can be grown as an annual or as a perennial crop, outside in fields or under protective covers, such as greenhouses. The ability to grow and produce a high-quality yield in such a wide range of climates has made peppers a common crop worldwide. To produce high yields of good quality, they do best with a long frost-free season. Peppers are highly susceptible to frost and grow poorly at temperatures of 5–15°C. They need relatively high soil temperature (25–31°C) for optimal germination and emergence.

Soil

The ideal soil for producing peppers, as for most crops, is a deep, well-drained, medium-textured loam or sandy loam that holds moisture and contains some organic matter. Most peppers are grown in soil of pH 7.0–8.5. Peppers are moderately sensitive to soil salinity.

Most peppers are grown in soil that is extensively tilled. No-tillage culture has been examined for pepper production, but peppers do not tolerate it as well as tobacco or tomato. The residual dead sod and residues of the previous crop are a potential source of disease organisms that may attack newly transplanted pepper plants before they become established. However, peppers can be produced in no-tillage culture with adequate pest and disease control.

In standard tillage methods, preparation of the soil involves plowing, deep chiseling, disking, smoothing, and listing. Laser leveling a field to a grade of 0.01 to 0.03% in one or both directions aids in draining the field of extra water, which in turn reduces the risk of root diseases.

Peppers can be grown in a level field or in raised beds. Raised beds are used in some areas to allow furrow irrigation and in others to ensure drainage. For direct-seeded crops, raised beds allow improved control of surface moisture, thereby reducing the chance of infection by soilborne organisms. Raised beds provide protection from root flooding by allowing the root zone to drain after heavy rains. The simplest way to form beds is by listing soil into ridges. The ridges are kept moist either with irrigation or rain to establish the field.

Planting

Peppers can be established in the field by direct seeding, by transplanting seedlings grown in multicellular trays in a greenhouse, or by transplanting bare-root seedlings grown in the field. Each method has advantages, and each is suitable for specific production systems. Direct seeding requires less labor and is less costly. However, with new hybrid cultivars costing 10 to 20 times more than open-pollinated ones, transplanting to a field stand may be the only economical option. Only the best-quality seed should be used to direct-seed a field or to start transplants in a greenhouse.

Transplanting

Most peppers planted today are started from transplants. A transplant is a plant moved from a culture medium to soil or from one soil or to another. Transplants are produced in greenhouses or hotbeds or (in mild climates) outdoor seedbeds six to eight weeks prior to field planting. Field seedbeds produce bare-root transplants, whereas peppers grown in greenhouses are containerized for transplanting.

Care should be taken in selecting a source of transplants, since they can carry diseases to production fields. Field-grown transplants have a greater likelihood of introducing soilborne pathogens. Greenhouse-grown transplants are more likely to be disease-free, because they are direct-seeded in a soilless growing medium. Soilless mixes provide good drainage and aeration but excellent water retention. Commercial mixes are free of pathogens, insects, and weed seeds. Whether seed is sown directly in the field or in the greenhouse, it can be treated with a fungicide to reduce damping-off. Foliar diseases, such as powdery mildew, bacterial spot, and some viral diseases, are potential problems in greenhouse-grown transplants.

Pepper plants are hardened off to reduce transplant shock about a week or two before they are moved to the field. Stem diameter is very important for the survival of transplants. The thicker the stem, the higher the survival rate. Some growers clip or pinch back transplants to three or more leaves before taking them to the field. Whenever plants are clipped, the likelihood of introducing pathogens increases. Growth can be delayed if young pepper plants are substantially overhardened.

Mulches

Plastic (polyethylene) mulches have been used for pepper culture since the early 1960s. They have been shown to moderate soil temperatures, hasten maturity, increase yields, increase fruit quality, and help control diseases. Plastic-mulched raised beds are commonly used for bell peppers and other fresh-market peppers. The mulch is clear, black, or coated. Coated mulch can be painted or tinted almost any color. The color of the mulch surface affects the growth and development of pepper plants. Black is the most common color. In areas where late summer or fall plantings are possible and soil warming is not beneficial, a white-surfaced mulch is often used. Pepper plants grown with red mulch are taller and heavier than plants grown with black, white, or yellow mulch. Plastic mulches are often used in conjunction with drip irrigation when pepper transplants are being established. A reduction of pepper disease has been reported when plastic mulch is used with soil fumigation and drip irrigation.

Mulches have also been shown to reduce aphid infestation and associated virus damage. Reflective mulch in particular reduces the incidence of virus disease in peppers and increases yield. Peppers grown with reflective silver plastic mulch produce greater yields than plants grown with black plastic mulch or on bare ground, even when infection by aphid-borne viruses is not apparent. The increased yields have been attributed to greater reflection of light (photosynthetically active radiation) from the aluminum-painted polyethylene.

Irrigation

Irrigation is not needed in areas with regular and ample rain, but it is essential for providing adequate moisture for the production of peppers in arid and semiarid regions and in the dry season in the humid tropics. Many hectares of peppers are grown with irrigation in many regions of the world. Peppers can require up to 60 to 75 ha-cm of water during a growing season. They are sensitive to moisture stress at flowering and fruit setting. If plant growth is slowed by moisture stress during blooming, blossoms and immature fruit are likely to drop. Blossom-end rot can occur in plants that are stressed when young fruit is developing rapidly.

Excess irrigation can be as harmful to a pepper crop as too little water. Phytophthora root rot can develop if water stands in a field for more than 12 hr when inoculum is present, so a means of draining the field is helpful. Frequent light irrigation is better for peppers than infrequent heavy irrigation, because of their shallow roots.

Drip irrigation is one method of supplying optimum water for pepper production and conserving water when rainfall is low. It also allows for frequent application of low levels of soluble nutrients to the root zone. Drip irrigation together with intensive cultural practices, such as mulching, generally results in yield increases.

Fertilization

A soil test of the field before planting and subsequent foliar testing during the growing season is beneficial to the health of a pepper crop. With information about soil nutrient levels before planting, a grower can adjust fertilizer application to provide optimum levels of nutrients while reducing fertilizer runoff.

Pepper benefits from nitrogen, but too much can overstimulate growth, resulting in large plants with few early fruits. During periods of high rainfall and humidity, excess nitrogen delays maturity, resulting in succulent late-maturing fruit and an increased risk of severe plant or fruit rots.

Inadequate amounts of several other nutrients during the growing period have been reported to cause pepper yield losses (see also Part V, Nutritional Disorders).

Phosphorus deficiency causes plants to be weak and produce narrow, glossy leaves, which turn grayish green. The red or purple coloration of stems and leaves often associated with phosphorus deficiency does not develop in pepper fruit. Vegetative tissues with a phosphorus content of 0.09% or lower exhibit deficiency symptoms.

Low levels of potassium cause a bronzing of pepper leaves, followed by necrosis and leaf drop. Symptoms of deficiency are associated with potassium levels of 1.17% or lower in vegetative growth.

Low calcium levels result in stunting of plants and severe blossom-end rot.

Magnesium deficiency in peppers is characterized by pale green leaves with interveinal yellowing, leaf drop, stunting of plants, and undersized fruit.

Peppers are sensitive to sodium, which can cause reduced yields and fruit weight.

Manganese toxicity may become apparent if the pH drops to 5 for a prolonged period. The symptoms are "burn" spots on leaves near the top of the plant.

Boron deficiency is expressed as a yellow discoloration of growing tips 30 cm below the top of the plant. Leaf veins turn brown, a symptom that is clearly visible when affected leaves are held up to the light. The deficiency is a result of poor root growth, because boron is taken up by the young root tips.

Flower Drop

Flower drop, or abscission (dropping) of flower buds, flowers, and immature fruit, is a serious problem in several pepper production regions. It occurs under various conditions: heat stress, insufficient water, treatment with growth regulators, and excessive or deficient levels of nutrients have been reported as causal agents. The best protection is to avoid overfertilizing and underwatering. When the disorder is corrected, plants will resume flowering and fruiting. Cultivars differ in their susceptibility to stress-induced flower drop.

Growth Regulators

Many growth regulators have been reported to affect peppers. The most common or most frequently studied are gibberellic acid and ethephon.

Young pepper plants treated with gibberellic acid (GA_3) before the initiation of floral organs subsequently form flowers with abnormalities.

Ethephon has been successfully used to concentrate red fruit maturity. However, it has had variable effects as a pepper fruit-ripening agent. Ethephon treatment often results in defoliation and fruit abscission, and it has been reported to be a possible cause of chlorosis. High temperatures after treatment have been found to accelerate fruit ripening, defoliation, and abscission, while low temperatures have reduced or negated the effects of the treatment. Multiple applications of ethephon at low concentrations may provide more consistent results than a single application at a higher concentration.

Greenhouse Production

The greenhouse is the ultimate climate modification for pepper production. It protects peppers from adverse climate and some pests and provides an elevated temperature year-round. Growing peppers in a greenhouse is similar to growing tomatoes, a crop that has been the subject of extensive greenhouse research.

Most greenhouse peppers are grown in a soilless medium. Hydroponic culture involves the production of peppers in sand, gravel, or an artificial soilless mix in bags, tubes, tubs, tanks, or troughs designed to allow the circulation of nutrient media needed for crop growth. It is estimated that 80% of greenhouse peppers are grown in soilless material, and the industry is moving toward 100%. In the Netherlands, all peppers are produced in rock wool.

Integrated pest management (IPM) is commonly practiced in greenhouse pepper production. IPM does not exclude the use of pesticides in the greenhouse. Rather, pesticides are used in combination with cultural, natural, mechanical, and biological control as well as insect monitoring to maximize the effectiveness of pest management. Reduced use of pesticides on more effective application schedules not only reduces their adverse effects on the environment and people but also lowers the risk of inducing resistance in pests and pathogens.

Seed Production

Pepper seed entering into interstate commerce in the United States must meet the requirements of the Federal Seed Act. Pepper seed packaged in large containers or packets must be labeled with the cultivar name and the word *hybrid,* if appropriate. In addition, the label should state the name of the shipper (producer), the germination rate, lot number, and any seed treatments. It is always wise to save a small portion of the seed. If problems arise later that can be ascribed to seed, examination of a seed sample may allow the cause to be identified.

The pepper seed is covered by a parchment-like seed coat. Like that of other solanaceous crops, the seed coat is derived from a single integument. It is usually smooth but in some cases is slightly rough and subscabrous. The seed coat does not impose any mechanical restriction on germination.

The seeds are straw yellow, tan, or black. As they age and lose viability, they can turn brown. Seed size depends on the variety and the growing conditions. Usually larger fruit has larger seeds. Most seeds fall in the range of 2.5–6.5 mm in length and 0.5–5.0 mm in width. A thousand seeds weigh 5–7 g. Seeds account for approximately 20% of the dry weight of pepper fruit.

There appear to be no special requirements for light in pepper seed germination. Fluorescent light sources used in germination cabinets do not inhibit germination, but nor do they promote it. Therefore, the presence or absence of light is not a factor in pepper seed germination.

Peppers have a prolonged germination period and an optimum germination temperature of about 30°C. The rate of germination and emergence is markedly reduced at temperatures in the range of 15–20°C.

Peppers are cross-pollinating plants, with recorded rates of outcrossing ranging from 2 to 90%. Outcrossing is associated with natural insect pollinators, not rain or wind. Seed should be purchased from a reputable source, to ensure that it is truly the seed of the cultivar that it is claimed to be. To produce large amounts of genetically pure seed, seed certification programs employ isolation as the control mechanism. The New Mexico Crop Improvement Association has established isolation requirements ranging from 1 mile for the Foundation class of seed to 1/4 mile for the Certified class.

Seed Disorders

Nutrient-deficient pepper plants can produce fewer seeds. Phosphorus nutrition of the parent plants does not seem to be important in influencing the performance of their seeds. However, potassium-deficient parent plants produce a high proportion of abnormal seeds with dark-colored embryos and seed coats. Both normal and abnormal seeds from such plants have a lower germination rate than control seeds, and their viability declines rapidly in storage.

A disorder called fish-mouth occurs when pepper seeds are harvested immaturely, before the endosperm has fully developed. The seed has a characteristic fish-mouth appearance, and hence the name.

Seed Treatments

Seed treatments are highly recommended, because viruses, fungi, and bacteria seeds may be carried on or inside the seed coat. Peppers germinate and emerge slowly, and therefore they are particularly susceptible to damping-off. Treating seed with a fungicide will help prevent seedling losses. Sodium hypochlorite treatment is a well-established procedure for reducing fungal and bacterial contamination of seeds and is effective in reducing the transmission of the bacterial spot pathogen. Trisodium phosphate is recommended as a treatment for viral contamination of pepper seed coats. Hot-water treatment is an old standard method that has been reported to control bacterial spot. The recommended treatment is 90°C for 25 min. An accurate thermometer is essential for hot-water treatment.

Selected References

Bosland, P. W. 2000. *Capsicum:* A Comprehensive Bibliography. 7th ed. Chile Institute, Las Cruces, N.M.

Bosland, P. W., and Votava, E. J. 1999. Peppers: Vegetable and Spice Capsicums. CAB International, Wallingford, U.K.

DeWitt, D., and Bosland, P. W. 1993. The Pepper Garden. Ten Speed Press, Berkeley, Calif.

DeWitt, D., and Bosland, P. W. 1996. Peppers of the World: An Identification Guide. Ten Speed Press, Berkeley, Calif.

Ludlam, J., ed. 1999. Chile Peppers. Handbook Series. Brooklyn Botanic Garden, New York.

(Prepared by P. W. Bosland)

Part I. Diseases Caused by Infectious Agents

Diseases Caused by Bacteria

Bacterial Canker

Bacterial canker, a persistent and important disease of tomatoes, has occasionally been reported in peppers. Symptoms on pepper leaves and fruit may be severe, sometimes resulting in economic loss. The disease was first reported in peppers in Israel in 1970. It has been reported more recently in California, Indiana, and Ohio and in São Paulo, Brazil.

Symptoms

Lesions on pepper leaves are initially small (less than 1 mm in diameter), raised, and white. The lesions enlarge, their centers may turn brown and become necrotic, and they may be surrounded with a pale white or chlorotic halo (Plate 1). Lesions on fruit are initially small, raised, and white, later enlarging and coalescing to form spots 1–3 mm in diameter. Fruit lesions are often surrounded by a white halo and appear similar to bird's-eye spots on tomato fruit. Severely infected fruit may fail to develop normally.

Fruit and leaf symptoms of bacterial canker are sometimes mistaken for those of bacterial spot, caused by *Xanthomonas* species.

In most reports of bacterial canker of pepper, stem symptoms, such as cankers and vascular discoloration, and wilting were not observed in the field or induced in inoculated plants in a greenhouse.

Causal Organism

Bacterial canker is caused by *Clavibacter michiganensis* (Smith) Davis et al. subsp. *michiganensis* (syn. *Corynebacterium michiganense*), a Gram-positive, motile, rod-shaped bacterium, measuring 1 × 0.5 μm. It does not produce spores, is nonlipolytic, slowly liquefies gelatin, and may hydrolyze starch weakly. Colonies are generally pale yellow, circular, entire, and butyrous on trypticase soy and nutrient agar media. Virulent strains induce a hypersensitive response in leaves of tobacco (*Nicotiana tabacum*) and four-o'clock (*Mirabilis jalapa*). Strains from pepper cause typical bacterial canker symptoms in tomato, including leaf and fruit spots, stem cankers, vascular discoloration, and wilting. Strain identity can be confirmed by fatty acid methyl ester (FAME) analysis, carbon substrate utilization pattern, enzyme-linked immunosorbent assay (ELISA) with monoclonal antibodies specific to *C. michiganensis* subsp. *michiganensis,* and polymerase chain reaction (PCR) assay with primers CMM-5 and CMM-6.

Disease Cycle

The bacterial canker pathogen is seedborne in tomatoes and peppers. Bacteria infest the seed coat both internally and externally. High relative humidity and temperatures of 25–30°C during the day and 20–23°C at night favor the disease. Young pepper leaves are more susceptible to infection than older leaves. In many pepper production areas, seedlings are produced in greenhouses prior to transplanting in the field. Conditions during transplant production (high temperature, relative humidity, and plant population density) are ideal for spread of the pathogen, even though symptoms are not always observed in the greenhouse. The pathogen enters plants through wounds or directly through stomata. It is spread to neighboring plants by splashing water, insects, tools, and worker contact. It can survive in or on tomato seed for five years (long-term survival on pepper seed has not been determined) and in soil associated with infested plant residues and weed hosts.

Epidemiology

There are several possible sources of inoculum: infested pepper seed, infected tomato seedlings in the greenhouse, infected tomatoes in production fields adjacent to pepper fields, infested plant residue, and weed hosts. The epidemiology of bacterial canker of pepper has not been fully documented. However, like bacterial canker of tomato, it is favored by high temperatures and relative humidity, rainfall, overhead irrigation, and conditions that result in significant wounding of leaves and fruit. In all reports to date, pepper strains have been fully compatible with tomatoes and have induced symptoms in tomato similar to those caused by tomato strains. Therefore, infected peppers may serve as a source of inoculum for tomatoes in the greenhouse and field.

Control

Bacterial canker is better prevented than controlled. Pathogen-free seed is the most effective means of managing this disease in both peppers and tomatoes. For peppers, hot-water treatment reduces or eliminates both the bacterial canker and bacterial spot pathogens in and on seeds. Other cultural methods include water management in the transplant production greenhouse, physical separation of tomato and pepper transplants, and good sanitation practices. In the field, copper bactericides are often applied but are minimally effective, at best. Peppers should be rotated out of all solanaceous crops for at least three years, and weeds should be controlled in planting areas.

Selected References

Almeida, I. M. G. de, Malavolta, V. A., Jr., and Robbs, C. F. 1996. Bacterial canker of pepper: Systemic infection with seed transmission. Summa Phytopathol. 22:112–115.

Dreier, J., Bermpohl, A., and Eichenlaub, R. 1995. Southern hybridization and PCR for specific detection of phytopathogenic *Clavibacter michiganensis* subsp. *michiganensis*. Phytopathology 85: 462–468.

Ivey, M. L. L., and Miller, S. A. 2000. First report of bacterial canker of pepper in Ohio. Plant Dis. 84:810.

Lai, M. 1976. Bacterial canker of bell pepper caused by *Corynebacterium michiganense*. Plant Dis. Rep. 60:339–342.

Latin, R., Tikhonova, I., and Rane, K. 1995. First report of bacterial canker of pepper in Indiana. Plant Dis. 79:860.

Volcani, Z., Zutra, D., and Cohn, R. 1970. A new leaf and fruit spot disease of pepper caused by *Corynebacterium michiganense*. Plant Dis. Rep. 54:804–806.

(Prepared by S. A. Miller)

Bacterial Spot

Bacterial spot is present in most pepper-growing regions, but it is most severe in tropical and subtropical regions with substantial rainfall. The disease is similar to bacterial spot of tomato, which was observed first on tomato in the United States and South Africa in 1912 and 1914, respectively. A similar disease of pepper was described in 1918. The causal agent isolated from tomato was characterized in South Africa in 1921. In 1922, it was determined that the organism infecting pepper was closely related to the one isolated from tomato, and in 1925 it was confirmed that the isolate from tomato and the isolate from pepper were the same organism.

The disease is most important in the southeastern United States, especially Florida, where it is a major problem. Crop losses result both from the actual yield reduction due to defoliation and from severe spotting of fruit, which renders it unfit for market.

Symptoms

The bacterial spot pathogens can infect all aboveground parts of the plant. Spots form on leaves, stems, and fruit, beginning as small, brown, water-soaked lesions, which turn brown and become necrotic in the center (Plate 2). The spots are water-soaked during rainy periods or when dew is present. They rarely enlarge to more than 3 mm in diameter.

Lesions are generally sunken on the upper surface, and leaf lesions are slightly raised on the lower surface. When conditions are optimal for disease development, leaf spots coalesce and form large blighted areas. A general yellowing of leaflets may occur following infection and often leads to premature leaf drop (Plate 3).

Fruit lesions begin as circular green spots. As the spots enlarge, reaching a diameter of 2 to 3 mm, they turn brown and acquire a cracked, roughened, wart-like appearance (Plate 4).

Causal Organisms

Bacterial spot is incited by two major groups of bacteria, *Xanthomonas campestris* pv. *vesicatoria* (Doidge) Dye and *X. vesicatoria* (ex Doidge) Vauterin et al. These bacteria are motile, strictly aerobic, Gram-negative rods, measuring 0.7–1.0 × 2.0–2.4 μm, and having a single polar flagellum. On nutrient agar they grow relatively slowly, forming colonies that are circular, wet, shining, Naples yellow, and entire in appearance. The bacteria produce acid but no gas from arabinose, glucose, sucrose, galactose, trehalose, cellobiose, and fructose. Both species produce xanthomonadins, the yellow pigments present in most *Xanthomonas* species.

Disease Cycle and Epidemiology

The bacterial spot pathogens are seedborne, being present within the seed or on its surface. Dissemination of bacteria on seed and transplants is an important means of dispersal. Temperatures between 24 and 30°C, high precipitation, and high relative humidity favor disease development. The pathogens are disseminated within a field by wind-driven rain droplets, clipping of transplants, and aerosols. They penetrate plants through stomates and wounds created by wind-driven sand, insect punctures, or mechanical means.

Control

Three single dominant genes conferring resistance to pathogenic *Xanthomonas* strains have been identified in *Capsicum*. The genes, designated *Bs1*, *Bs2*, and *Bs3*, were first found in PI 163192 (*C. annuum* L.), PI 260435 (*C. chacoense* Hunz.), and PI 271322 (*C. annuum*), respectively. None of these genes confers resistance to all strains of *Xanthomonas* pathogenic in pepper. Recently, another source of resistance was found in PI 235047 (*C. pubescens* Ruiz & Pav.). Eleven pathogenic races have been distinguished in studies of near-isogenic lines of the cultivar Early Calwonder containing one of the three single resistance genes and the resistance derived from *C. pubescens*. Resistant varieties containing combinations of the *Bs1*, *Bs2*, and *Bs3* genes are available.

Rotation of fields in an attempt to avoid carryover of inoculum on volunteers and crop residue is an important control measure. Peppers should never follow tomatoes, and tomatoes should never follow peppers.

Transplants should be free of disease. To facilitate the production of disease-free transplants, they should be raised in an area separate from field production of tomatoes and peppers.

Seed treatment should be used to reduce possible transmission of the pathogens.

Cull piles should not be located near greenhouses or field operations.

Bactericides or fungicide-bactericide combinations may be applied where recommended.

Selected References

Bonas, U., Stall, R. E., and Staskawicz, B. 1989. Genetic and structural characterization of the avirulence gene avrBs3 from *Xanthomonas campestris* pv. *vesicatoria*. Mol. Gen. Genet. 218:127–136.

Cook, A. A., and Guevara, Y. G. 1984. Hypersensitivity in *Capsicum chacoense* to race 1 of the bacterial spot pathogen of pepper. Plant Dis. 68:329–330.

Cook, A. A., and Stall, R. E. 1963. Inheritance of resistance in pepper to bacterial spot. Phytopathology 53:1060–1062.

Doidge, E. M. 1921. A tomato canker. Ann. Appl. Biol. 7:407–430.

Gardner, M. W., and Kendrick, J. B. 1923. Bacterial spot of tomato and pepper. Phytopathology 13:307–315.

Higgins, B. B. 1922. The bacterial spot of pepper. Phytopathology 12: 501–516.

Kim, B.-S., and Hartmann, R. W. 1985. Inheritance of a gene (Bs$_3$) conferring hypersensitive resistance to *Xanthomonas campestris* pv. *vesicatoria* in pepper (*Capsicum annuum*). Plant Dis. 69:233–235.

Kousik, C. S., and Ritchie, D. F. 1996. Race shift in *Xanthomonas campestris* pv. *vesicatoria* within a season in field-grown pepper. Phytopathology 86:952–958.

McInnes, T. B., Gitaitis, R. D., McCarter, S. M., Jaworski, C. A., and Phatak, S. C. 1988. Airborne dispersal of bacteria in tomato and pepper transplant fields. Plant Dis. 72:575–579.

Minsavage, G. V., Dahlbeck, D., Whalen, M. C., Kearney, B., Bonas, U., Staskawicz, B., and Stall, R. E. 1990. Gene-for-gene relationships specifying disease resistance in *Xanthomonas campestris* pv. *vesicatoria*–pepper interactions. Mol. Plant-Microbe Interact. 3: 41–47.

Peterson, G. H. 1963. Survival of *Xanthomonas vesicatoria* in soil and diseased tomato plants. Phytopathology 53:765–767.

Pohronezny, K., Stall, R. E., Canteros, B. I., Kegley, M., Datnoff, L. E., and Subramanya, R. 1992. Sudden shift in the prevalent race of *Xanthomonas campestris* pv. *vesicatoria* in pepper fields in southern Florida. Plant Dis. 76:118–120.

Sahin, F., and Miller, S. A. 1995. First report of pepper race 6 of *Xanthomonas campestris* pv. *vesicatoria,* causal agent of bacterial spot of pepper. Plant Dis. 79:1188.

Sahin, F., and Miller, S. A. 1998. Resistance in *Capsicum pubescens* to *Xanthomonas campestris* pv. *vesicatoria* pepper race 6. Plant Dis. 82:794–799.

Sherbakoff, C. D. 1918. Report of the associate plant pathologist. Fla. Agric. Exp. Stn. Rep. 1916–1917:66R–86R.

(Prepared by J. B. Jones and K. Pernezny)

Bacterial Wilt

Bacterial wilt is a serious soilborne disease of many economically important crops, including tomato, potato, tobacco, banana, and eggplant. Peppers, however, are not as susceptible as those crops, and economic losses of peppers usually occur only under certain climatic conditions in tropical and subtropical regions. The disease has gained importance in protected cultivation (plastic tunnel or greenhouse), where temperatures are usually higher and crop rotation is not properly performed for economic reasons. In the southeastern United States, bacterial wilt of pepper is not a serious disease, apparently because prevalent strains from tomato and tobacco do not cause wilt symptoms in pepper. Efforts have been made to identify sources of resistance in peppers in Japan, Taiwan, and Brazil.

Symptoms

Diseased plants can be found scattered in the field, but bacterial wilt usually occurs in foci associated with the accumulation of water in low areas. In furrow-irrigated crops, it is common to find wilted plants in contiguous sections of rows, as a result of the spread of inoculum through the water channel. The initial symptom in mature plants under natural conditions is similar to that observed in tomato and potato: wilting of upper leaves on hot days followed by recovery throughout the evening and early hours of the morning. The wilted leaves maintain their green color and do not fall as the disease progresses. Under conditions favorable for the disease, entire plants wilt (Plate 5). A dark brown discoloration appears in vascular tissues in the lower stems of wilted plants (Plate 6).

These symptoms are very similar to those of Phytophthora blight, induced by *Phytophthora capsici*. However, an extensive external darkening of the lower stem is not often observed in plants with bacterial wilt. Furthermore, the stem of a plant with bacterial wilt, cut in cross section, exudes a white, milky strand of bacterial cells in clear water (Plate 7). This feature clearly distinguishes bacterial wilt from wilting caused by fungal or oomycete pathogens.

Causal Organism

Bacterial wilt is caused by *Ralstonia solanacearum* (Smith) Yabuuchi et al. This species was known for many years as *Pseudomonas solanacearum* (Smith) Smith. The genus *Ralstonia* was established to accommodate *R. solanacearum* and the closely related species in rRNA homology group II, *R. pickettii* and *R. eutropha*. *R. solanacearum* is now widely accepted as the name of the bacterial wilt pathogen.

R. solanacearum is a Gram-negative rod (0.5–0.7 × 1.5–2.5 μm). It is motile, aerobic, and oxidase- and catalase-positive, and it accumulates poly-β-hydroxybutyrate. Colonies are nonfluorescent on complex media. Most strains, with the exception of biovar 2 strains, produce nitrite from nitrate. The bacterium grows in 0.5 and 1% but not 2% NaCl broth. In general, virulent colonies of *R. solanacearum* are fluidal and irregular, with a characteristic light pink center and whitish periphery on tri-

phenyl tetrazolium chloride (TTC) medium. Avirulent mutants lacking the production of extracellular polysaccharide appear small, round, red, and butyrous. TTC medium is easy to prepare, and colonies grown on it are consistent in appearance, so that it the most widely used medium for distinguishing virulent *R. solanacearum* colonies in pure culture. The bacterium can be stored in sterile distilled water at room temperature for several months without significant loss of virulence. However, the organism is best stored at –80°C in an aqueous glycerol suspension.

R. solanacearum is a complex and diverse species. It is a pathogen of several hundred plant species belonging to over 50 families. At the subspecies level, the pathogen was divided by I. Buddenhagen and co-workers (1962) into three races, distinguished by their ability to infect different plant species. Race 1 has a very wide host range and is known as the solanaceous race. Race 2 is the *Musa* race. Race 3 is the potato race. Later, races 4 (the mulberry race) and 5 (the ginger race) were proposed. The grouping by races to differentiate strains of *R. solanacearum* has become less practical with the rapid discovery of new host plants. Strains of *R. solanacearum* have also been found to differ in their biochemical characteristics. A. C. Hayward proposed dividing the species into different biovars, distinguished by their ability to utilize or oxidize three hexose alcohols and three disaccharides. Recent molecular techniques will certainly be an important aid to better understand the diversity *R. solanacearum*. On the basis of restriction fragment length polymorphism (RFLP) analysis, the species has been divided into two broad RFLP groups: division I accommodates strains of biovars 3, 4, and 5, and division II contains biovars 1 and 2.

Disease Cycle and Epidemiology

Bacterial wilt of pepper is caused predominantly by biovars 1 and 3 of *R. solanacearum*. As members of race 1, these biovars have wide host ranges, which guarantee their long-term survival in soil in the absence of a susceptible crop. The pathogen can survive in the rhizosphere of nonhost plants, including weeds. Soil factors also influence the survival of the bacterium. For example, bacterial wilt is an important disease of tomato in north Florida, but it rarely occurs in calcareous soils with high pH. Moderate pH and moderate to high temperatures are associated with the survival of the bacterium in soil and greater disease severity. Soils suppressive of bacterial wilt promote desiccation of the pathogen and antagonistic microbial activity.

Infested soil is the main source of inoculum. It is not unusual to find bacterial wilt in the first crop in recently cleared land in tropical and subtropical regions. Disease-free areas can be infested by infected planting material (tomato or pepper transplants or potato tubers), contaminated water or machinery, and laborers who carry bacteria from infested fields. Transmission by seed is not considered important in pepper, even though *R. solanacearum* has been reported to be seed-transmitted in tomato, eggplant, and peanut. Epiphytic colonization of the true leaves of pepper has been shown to occur at high relative humidity. In the field, the most common means of dissemination of the pathogen is contaminated irrigation water.

R. solanacearum can infect undisturbed roots of susceptible hosts by entering microscopic wounds caused by the emergence of lateral roots. Transplanting, nematodes, insects, and agricultural equipment are other common causes of root wounds, which allow the bacterium to enter the plant. It then colonizes the cortex and makes its way toward the xylem vessels, from which it rapidly spreads in the plant. Bacterial masses prevent the flow of water from the roots to the leaves, and the plants wilts as a result. The severity of the disease depends on soil temperature, soil moisture, soil type (which influences soil moisture and microbial populations), host susceptibility, and the virulence of the strain of the pathogen. High temperature

(30–35°C) and high soil moisture are the main factors associated with high incidence and severity of bacterial wilt. Under these conditions, large populations of bacteria are released into the soil from the roots as the plant wilts.

Control

Because it is caused by a soilborne pathogen with a wide host range, bacterial wilt is very difficult to control after it has been established in a field. No single measure totally prevents losses caused by the disease. Quarantine regulations must be enforced to prevent the introduction of exotic strains of the pathogen that may attack pepper.

Cultural practices can reduce the incidence and severity of bacterial wilt, allowing the disease to be manageable. Seedlings should be free from infection by *R. solanacearum*. It is mandatory that commercial seedling producers use irrigation water not contaminated with the pathogen. Fields should not be irrigated excessively, because soil moisture favors a buildup of the pathogen. Crop rotation with nonsusceptible crops reduces soilborne populations of the bacterium. The rotation period and the choice of rotational crops may be limited by economic considerations, but grasses are always preferred. Shifting planting dates to cooler periods of the year can be very effective in reducing disease development. Soil amendments with inorganic and organic mixtures reduce wilt incidence in some locations, but more research is required to develop economically feasible materials and doses and explicit recommendations for treatment.

In Brazil and Taiwan, genotype MC-4 of *Capsicum annuum* from Malaysia has been found to be very resistant to various strains of biovars 1 and 3 and is recommended for breeding programs. In Japan, India, and Taiwan, other sources of resistance have been identified in sweet bell pepper. Since tomato and potato cultivars resistant to bacterial wilt have failed across locations because of variation in environmental factors and strains, resistance should be seen as one component of an integrated program for management of this disease.

Selected References

Buddenhagen, I., Sequeira, L., and Kelman, A. 1962. Designation of races in *Pseudomonas solanacearum*. (Abstr.) Phytopathology 52:726.

Denny, T. P., and Hayward, A. C. 2001. *Ralstonia*. Pages 151–174 in: Laboratory Guide for Identification of Plant Pathogenic Bacteria. 3rd ed. N. W. Schaad, J. B. Jones, and W. Chun, eds. American Phytopathological Society, St. Paul, Minn.

Elphinstone, J. G., Hennessy, J., Wilson, J. K., and Stead, D. E. 1996. Sensitivity of different methods for the detection of *Pseudomonas solanacearum* (Smith) Smith in potato tuber extracts. Bull. OEPP/EPPO Bull. 26:663–678.

Granada, G. A., and Sequeira, L. 1983. Survival of *Pseudomonas solanacearum* in soil, rhizosphere, and plant roots. Can. J. Microbiol. 29:443–440.

Hayward, A. C. 1964. Characteristics of *Pseudomonas solanacearum*. J. Appl. Bacteriol. 27:265–277.

Hayward, A. C., and Hartman, G. L., eds. 1994. Bacterial Wilt: The Disease and Its Causative Agent, *Pseudomonas solanacearum*. CAB International, Wallingford, U.K.

Kelman, A. 1954. The relationship of pathogenicity in *Pseudomonas solanacearum* to colony appearance on a tetrazolium medium. Phytopathology 44:693–695.

Pradhanang, P. M., Elphinstone, J. G., and Fox, R. T. V. 2000. Identification of crop and weed hosts of *Ralstonia solanacearum* biovar 2 in the hills of Nepal. Plant Pathol. 49:403–413.

Quezado-Soares, A. M., and Lopes, C. A. 1995. Stability of the resistance to bacterial wilt of the sweet pepper 'MC-4' challenged with strains of *Pseudomonas solanacearum*. Fitopatol. Bras. 20:638–641.

(Prepared by M. T. Momol,
P. M. Pradhanang, and C. A. Lopes)

Syringae Seedling Blight and Leaf Spot

Strains of *Pseudomonas syringae* have been isolated from diseased pepper plants exhibiting a seedling blight or leaf spot. Most reports of the disease are from southern and southeastern Europe (Hungary, Yugoslavia, Macedonia, and Italy). It appears to be particularly important in pepper transplant production under cool conditions and high humidity. Young plants in hotbeds or plastic tunnels and young leaves of mature plants in the field appear to be more susceptible than older leaf tissue. The disease is not important in pepper production in the United States at this time, having been reported only once in the United States, in North Carolina in 1964.

Symptoms

The initial symptoms of seedling blight and leaf spot are observed on pepper seedlings grown in hotbeds or plastic tunnels in nurseries. Lesions are initially water-soaked and later turn dark brown to black. Necrotic spots may develop on cotyledons and the first true leaves (Plate 8). The spots do not have a chlorotic halo. They are initially punctiform but tend to coalesce, forming larger, irregularly shaped lesions, often located along the cotyledon edge or covering the entire cotyledon and causing blight. If apical tissue is affected, plant growth stops. Under conditions favorable for the disease, water-soaking and necrosis spread along the hypocotyls, resulting in rapid collapse and death of the plants. These symptoms are frequently misidentified as damping-off, which is caused by several soilborne pathogens. Lesions that form on the first true leaves cause them to be misshapen. Leaf epinasty can occur.

In the field, necrotic spots 1–3 mm in diameter surrounded by a yellow halo develop on leaves (Plate 9). The symptoms usually occur in the spring but may also be observed during the summer when temperatures are below normal (16–24°C) and there is frequent rain. When conditions are favorable for disease development, spots coalesce, killing large areas of leaf tissue and causing premature leaf abscission. Warm, dry weather slows disease development, and plants recover; however, affected leaves remain deformed, with cracks in necrotic areas.

Fruit infection has not been observed. As a consequence of defoliation, the quality of the fruit and total yield are substantially decreased.

Causal Organism

In early studies *Pseudomonas syringae* was identified as the cause of seedling blight and leaf spot of pepper. Recent reports identify *P. syringae* pv. *syringae* van Hall as the causal agent. The bacterium is non-spore-forming, Gram-negative, and rod-shaped. It produces green fluorescent pigment on King's medium B, forms levan on 5% sucrose medium, and induces a hypersensitive reaction in tobacco. Strains isolated from pepper are negative for oxidase and arginine dihydrolase activity and for soft rot of potato. These characteristics help identify the bacterium as a *P. syringae* pathovar. The bacterium is aerobic, does not grow at 41°C, liquefies gelatin, does not reduce nitrates, and does not hydrolyze starch. Pepper strains of the bacterium are positive for erythritol utilization, ice nucleation activity, and syringomycin production. On the basis of these characteristics, the pathovar has been classified as *P. syringae* pv. *syringae*. Only strains isolated from pepper induce typical symptoms in pepper plants.

Disease Cycle and Epidemiology

The source of inoculum is unknown. Air temperatures of 16–24°C and high humidity favor disease development. Within a plant bed or field, the pathogen spreads in splashing water from overhead irrigation, wind-driven rain droplets, or aerosols. It penetrates leaf tissue by entering natural openings or wounds.

Control

Facilities for transplant production should be disinfested or rotated away from pepper and related plant species for at least two successive years. Low temperatures and high humidity in plant nurseries should be avoided. Growers should use disease-free transplants. Crop rotation with a nonhost is recommended. Overhead irrigation of the crop should be avoided where the disease occurs. Appropriate bactericides can be applied according to recommendations.

Selected References

Arsenijević, M., and Balaz, J. 1978. Etiological studies of bacterial spot on pepper leaves. Contemp. Agric. (Novi Sad) 7–8:75–86.

Arsenijević, M., and Balaz, J. 1983. Bacterial leaf spot of pepper. Plant Prot. (Belgrade) 34(163):163–168.

Arsenijević, M., and Obradović, A. 1997. A pathovar of *Pseudomonas syringae* causal agent of bacterial leaf spot and blight of pepper transplants. Pages 61–66 in: Developments in Plant Pathology. Vol. 9, *Pseudomonas syringae* Pathovars and Related Pathogens. K. Rudolph, T. J. Burr, J. W. Mansfield, D. Stead, A. Vivian, and J. von Kietzell, eds. Kluwer Academic Publishers, Dordrecht, Netherlands.

Buonaurio, R., and Scortichini, M. 1994. *Pseudomonas syringae* pv. *syringae* on pepper seedlings in Italy. Plant Pathol. 43:216–219.

El-Kady, S., Hevesi, M., Sule, S., and Klement, Z. 1985. Pathological and serological separation of some *Pseudomonas syringae* strains isolated from apricot, sour-cherry and pepper hosts. Acta Hortic. 192:193–203.

Mitrev, S., Gardan, L., and Samson, R. 2000. Characterization of bacterial strains of *Pseudomonas syringae* pv. *syringae* isolated from pepper leaf spot in Macedonia. J. Plant Pathol. 3:227–231.

Obradović, A., Arsenijević, M., Marinković, N., and Mijatović, M. 1994. Bacterial spot of pepper transplants. Page 65 in: Book of Abstracts, Yugosl. Congr. Plant Prot., 3rd.

Obradović, A., Mavridis, A., Rudolph, K., Arsenijević, M., and Mijatović, M. 2000. Bacterial diseases of pepper in Yugoslavia. Proc. Int. Conf. Plant Pathogenic Bact., 10th. S. K. DeBoer, ed. Kluwer Academic Publishers, Dordrecht, Netherlands.

Person, L. H. 1964. A bacterial leafspot of pepper caused by *Pseudomonas syringae*. Plant Dis. Rep. 48:750–753.

(Prepared by A. Obradović)

Diseases Caused by Fungi and Oomycetes

Anthracnose

Anthracnose is usually a disease of maturing hot and sweet pepper fruit, but immature fruit and even leaves and stems may become diseased under the right circumstances. Severe losses can occur when the disease is left untreated and the weather is favorable for disease development. Pepper anthracnose occurs in all of the major vegetable-growing areas of the world.

Symptoms

Fruit lesions usually develop during ripening, but lesions can form on fruit of any size and on foliage and stems at later stages of infection. As the fruit ripens, susceptibility to infection increases. Fruit lesions are circular (Plate 10) and may reach a diameter of 30 mm or more on large fruit. At the lesion center are concentric rings, which are tan to orange to black, depending on the age of the lesion and the species of fungus causing the disease. The initial symptoms are indefinite tan lesions, which may appear a few days after infection. Acervuli containing salmon-colored spore masses often become visible (Plate 11). As the lesions grow, concentric black rings of fungal microsclerotia may be formed. Dark setae may be present. The lesions usually remain discrete, even as the number of lesions per fruit increases (Plate 12). Lesions may also appear on stems and leaves as irregularly shaped gray brown spots with dark brown edges. The spots are usually overlooked, but they can serve as a source of secondary inoculum.

Causal Organisms

Anthracnose of pepper is caused by several species of the genus *Colletotrichum,* including *C. gloeosporioides* (Penz.) Penz. & Sacc. in Penz. (teleomorph *Glomerella cingulata* (Stoneman) Spauld. & H. Schrenk), *C. capsici* (Syd.) E. J. Butler & Bisby, *C. coccodes* (Wallr.) S. J. Hughes, and possibly others. The taxonomy of *Colletotrichum* is in a state of flux. It is complicated by the wide range of phenotypic variability in these species and by the broad host range of *C. gloeosporioides.* More than one *Colletotrichum* species or biotype may be associated with a single host.

C. gloeosporioides forms one-celled, hyaline conidia, which are cylindrical with obtuse ends and measure 2.7–5.0 × 11.1–18.5 μm. It sometimes produces setae, but not always.

C. capsici forms conidia measuring 2.7–4.8 × 19.8–28.3 μm. This species does not form sclerotia, but it produces abundant setae.

C. coccodes forms conidia measuring 2.5–4.8 × 11.5–25.0 μm. They are massed in disk-shaped acervuli (fruiting bodies), which may contain black, septate setae, 65–112 μm long, although setae are not always formed. This species produces microsclerotia about 0.5 mm in diameter.

Disease Cycle and Epidemiology

Acervuli and sclerotia of anthracnose fungi have been identified in seed. Sclerotia may persist in plant debris from infected crops and in other plant species. They produce conidia, which can be dispersed in splashing water and thus spread to pepper foliage and fruit. Conidia germinate and form adhesive appressoria, which serve as survival structures until an infection peg penetrates the surface. Infection depends primarily on free moisture, and an extended period of leaf wetness is required. The longer the period of high relative humidity and rain, the greater the disease severity. Diseased fruit, foliage, and stems are sources of secondary inoculum, which spreads from plant to plant in the field. The dispersal of microsclerotia and conidia is favored by splashing rain and overhead irrigation. Optimum temperatures for infection are 20–24°C, but infection can occur at temperatures from 10 to 30°C.

Control

Planting pathogen-free seed and crop rotation are the most effective anthracnose management strategies. Seed can be disinfested with a 30-min soak in water at 52°C. Peppers should be rotated out of infested fields, which should be planted with nonsolanaceous crops for three years. Another disease management strategy is sanitation, such as deep plowing of crop residues and removal of plant material from the field.

Registered fungicides are available and may provide control of anthracnose when environmental conditions are less than optimal for disease development or when a low level of inoculum is present.

Selected References

Boucher, T. J., and Ashley, R. A., eds. 2000. Northeast Pepper Integrated Pest Management Manual. University of Connecticut Cooperative Extension, Storrs.

Freeman, S., Katan, T., and Shabi, E. 1998. Characterization of *Colletotrichum* species responsible for anthracnose diseases of various fruits. Plant Dis. 82:596–605.

Manandhar, J. B., Hartman, G. L., and Wang, T. C. 1995. Anthracnose development on pepper fruits inoculated with *Colletotrichum gloeosporioides*. Plant Dis. 79:380–383.

Manandhar, J. B., Hartman, G. L., and Wang, T. C. 1995. Conidial germination and appressorial formation of *Colletotrichum capsici* and *C. gloeosporioides* isolates from pepper. Plant Dis. 79:361–366.

Oh, I. S., In, M. S., Woo, I. S., Lee, S. K., and Yu, S. H. 1988. Anthracnose of pepper seedling caused by *Colletotrichum coccodes* (Wallr.) Hughes. Korean J. Mycol. 16:151–156.

(Prepared by S. A. Alexander and K. Pernezny)

Cercospora Leaf Spot (Frogeye Leaf Spot)

Cercospora leaf spot, also known as frogeye leaf spot, occurs in pepper in tropical and subtropical areas in many countries in Africa, Asia, South America (Venezuela), and the West Indies (Cuba, Jamaica, and Trinidad). In the United States, the disease is most common in the southeastern states, from Florida to North Carolina to Kentucky, and in Texas. It has also occurred, less commonly, in New Mexico, Arizona, and California. In Florida, the disease has been seen throughout the state, but it is most common in the northern part of the state at the end of the commercial harvesting season and in home gardens where peppers are grown during the summer. The pathogen is most active under hot conditions, which are typical in northern Florida during the summer, where nighttime temperatures are near 21°C and daytime temperatures are 30 to 35°C. It is not common in southern Florida, where peppers are grown commercially during the fall, winter, and spring, when the weather is cooler. Because frogeye leaf spot is not a major problem, very little research on the pathogen or the disease has been reported.

Symptoms

Lesions form on leaves, petioles, peduncles, and stems. Because of reduced photosynthesis and defoliation, fruit size may be reduced. The presence of a few strategically placed lesions, or even one lesion, on a leaf or petiole is enough to cause the leaf to abscise. The first symptoms are small, round, water-soaked spots, which develop into lesions with a light brown to white interior a brown to red to purple border (Plate 13). As the lesions expand, an outer water-soaked area and dark ring may form beyond the original border, so that the lesion center is surrounded with concentric rings. The lesions vary in size, ranging from 0.3 to 1.3 cm in diameter. The inner part of the lesion becomes brittle and cracks as it turns light in color. Stem lesions may become elongated, and a stem can be girdled by a single lesion. Conidia and conidiophores of the pathogen are formed in leaf spots and can be seen with the aid of a hand lens or microscope.

Causal Organism

Cercospora capsici Marchal & Steyaert causes frogeye leaf spot of pepper. A teleomorphic stage of this fungus is not known. *C. capsici* forms conidia on unbranched conidiophores, which are straight to slightly curved, pale brown, and 30–80 μm long and 4–6 μm wide. Conidial scars are clearly evident on the conidiophores after conidia have been dislodged. The conidia are several-celled, straight to slightly curved, hyaline, smooth, and 40–135 μm long and 3–5 μm wide. They are widest at the base and have a thickened scar at the point of attachment to the conidiophore.

Disease Cycle

Infection of host tissue occurs by the penetration of stomates by hyphae produced from germinating spores. In seven to 10 days, new spores are formed in the light-colored portion of the lesion. The spores are disseminated by wind, rain, irrigation, or mechanical means. The site of survival of *C. capsici* between crops is not known. However, *Cercospora* spp. are well known to carry over from one season to another in association with crop debris. *Capsicum annuum* and *C. frutescens* are the only known hosts of the pathogen.

Control

Frogeye leaf spot is typically a minor disease, and control measures are usually not needed. However, growers should be observant for any deviation from this pattern. Burying crop debris should be helpful. Crop rotation is known to reduce the early onset of other diseases caused by *Cercospora* spp. Where control is needed, several fungicides currently registered for use on pepper will reduce the disease if a spray program is initiated early in an epidemic.

Several cultivars have been identified as having some degree of resistance to *C. capsici*, but new varieties are routinely being developed, and there is a lack of adequate information on the variability in pathogenicity of different isolates of the fungus.

Some have speculated that *C. capsici* may be associated with pepper seed, but a clear association with seed has not been established.

Selected References

Kirk, P. M. 1982. *Cercospora capsici.* Descriptions of Pathogenic Fungi and Bacteria, no. 723. Commonwealth Mycological Institute, Kew, England.

Sherf, A. F., and MacNab, A. A. 1986. Vegetable Diseases and Their Control. 2nd ed. John Wiley & Sons, New York.

Weber, G. F. 1932. Diseases of peppers in Florida. Univ. Fla. Agric. Exp. Stn. Bull. 244.

(Prepared by T. A. Kucharek)

Charcoal Rot

Charcoal rot, caused by *Macrophomina phaseolina* (Tassi) Goidanich, is a disease of pepper at very high temperatures (32°C and above). *M. phaseolina* is a polyphagous species, attacking many crops, especially in the tropics and warm temperate areas. Pepper seems to be less susceptible to this pathogen than several other hosts, particularly legumes.

A characteristic symptom of charcoal rot is black, sunken cankers that form below the cotyledon node soon after emergence. Plants that become infected after maturity are stunted, become chlorotic in older leaves, and may wilt. The presence of many small, black sclerotia (0.5 mm in diameter) can give the interior of infected stems a grayish cast.

M. phaseolina produces pycnidia in diseased stems. The pycnidia are membranous to subcarbonaceous, erumpent, and globose and have inconspicuous, truncate ostioles. Conidiophores are simple, rod-shaped, and 10–15 μm long. Conidia are one-celled, hyaline, and elliptical to oval, measuring 16–29 × 6–9 μm.

Broad-spectrum fumigants can provide excellent control of the disease. Even in the absence of fumigants, charcoal rot does not seem to be a major disease of pepper at this time.

Selected Reference

Luttrell, E. S. 1946. A pycnidial strain of *Macrophomina phaseoli.* Phytopathology 36:978–980.

(Prepared by K. Pernezny)

Choanephora Blight (Wet Rot)

Choanephora blight of pepper is not a common disease worldwide. Although the disease was known in the 1800s, the first report associating it with the pathogen, *Choanephora cucurbitarum,* was published in 1903. The first description of the disease in pepper in India, in 1920, used the term *wet rot* to describe the general appearance of affected plants. The disease has also been called bud rot, blossom blight, blossom rot, and blossom mold. At the time of this writing, no approved name has been established for this disease in pepper. The name *Choanephora blight* will be used herein, because it is similar to the approved names of similar diseases caused by *C. curcurbitarum* in other crops (e.g., Choanephora fruit rot of squash).

Choanephora blights have occurred around the world in numerous plant species, including *Amaranthus* spp., bean (*Phaseolus vulgaris*), beet, *Chenopodium ambrosioides, Colocasia antiquorum,* cotton, cucurbits (many species), eggplant, guar, *Hibiscus* spp., okra, papaya, pea (*Pisum sativum*), peanut, pepper, *Petunia* ×*hybrida,* poinsettia, southern pea (*Vigna* spp.), soybean, sunn hemp, sweet potato, and zinnia. The pathogen has been isolated from soil, dead insects, and many crop species that are not generally thought to be susceptible to infection by this fungus (e.g., species of *Pinus, Citrus,* and *Gladiolus*). Collectively, Choanephora blights occur as diseases of leaves, fruit, blossoms, stems, and petioles.

C. cucurbitarum grows best in a warm environment with a high level of moisture. Choanephora blight of pepper has occurred in the United States, most commonly in Florida, since 1932. It has also been reported in Georgia and North Carolina and in northern states (Wisconsin, Ohio, and New York). The disease has been reported to be a persistent problem in India.

Diseases caused by *Phytophthora capsici* and *Botrytis cinerea* produce similar symptoms and thus have been confused with Choanephora blight in some field situations.

Symptoms

In Florida and India, Choanephora blight of pepper occurs mostly during periods of high relative humidity and in crops planted in the late summer to early fall, when the weather is warm. Peppers are most susceptible from the seedling stage to the early flowering stage, before woody tissue begins to develop in the stem. Individual branches of a plant may be infected and die back. In Florida, Choanephora blight is not common in crops planted in the spring, when temperatures at planting time are lower. Temperatures below 14°C strongly inhibit the disease. Infection of flowers, calyxes, petioles, young stems, and leaves results in continuous growth of the fungus over and in young succulent tissue. Individual leaves or entire plants may wilt (Plate 14) and die. Flowers or flower buds turn dark and wilt. Young fruit can be infected.

Soon after the first symptoms occur, while conditions are warm and moist, sporulation of *C. cucurbitarum* can easily be seen with the aid of a hand lens. The fungus appears as a silvery to metallic fuzz on the surface of the tissue. At the tips of individual silvery fungal strands are dark, knoblike objects, which contain reproductive structures (usually sporangiola). Morning is the best time to see signs of the fungus.

Causal Organism

Choanephora cucurbitarum (Berk. & Ravenel) Thaxt. is a fungus in the phylum Zygomycota, class Zygomycetes, order Mucorales, and family Choanephoraceae. It produces several structures that can serve as sources of inoculum, survival structures, or both. The most common reproductive structure of the fungus on host tissue is the sporangiolum, a modified sporangium, which is often called a conidium. Sporangiola are 8–30 μm long and 5–18 μm wide. They can germinate in $1\frac{1}{2}$ to 4 hr, bearing sporangiophores (conidiophores), which can be 10 mm high. The optimum temperature for the production of sporangiola is 25°C, but they can form at temperatures up to 31°C. Alternate light and dark (or low light) regimes enhance the production of sporangiola, particularly during the dark periods, and bright light inhibits it. Sporangiola can also form on culture media, but production is minimal in cultures that are not adequately aerated.

C. cucurbitarum forms sporangia, but they are not likely to be seen in the field. They have been produced in culture media along with sporangiola. The presence of sporangia on rotting pepper plants and *Colocasia antiquorum* has been reported in India. The optimum temperature for sporangial development is between 32 and 34°C. Sporangia range from 25 to 200 μm in diameter. They are initially white and then turn yellow and various shades of brown, but they appear black when viewed by reflected light. Sporangiospores are 12–48 μm long and 7–20 μm wide.

C. cucurbitarum is heterothallic, and therefore two mating types are required for the production of zygospores by the fungus. Zygospores have been formed in culture media but have rarely been seen in host tissue. They are small, ranging from 35 to 105 μm in diameter, and translucent reddish brown. The formation of zygospores is enhanced by continuous darkness, continuous light, extremes of temperature, and accumulation of CO_2 due to poor aeration. Germination of zygospores has been demonstrated once, but some researchers have been unable to observe it.

Chlamydospores are formed in chains within hyphae in culture and are potentially a survival structure of *C. cucurbitarum.* Germination of chlamydospores has been observed once.

Disease Cycle

C. curcurbitarum produces several types of spores in culture, but sporangiola appear to be the primary inoculum. The fungus can survive in soil and subsequently infect host tissue. It has a positive chemotropic response to flower parts, to a greater degree than other plant parts, which may partially explain why Choanephora blight is typically associated with flowers of crops such as peppers and squash. Bees and beetles that colonize flowers have been shown to transmit viable inoculum of *C. cucurbitarum.* Symptoms are the result of exuded enzymes (e.g., pectin-methylesterase and pectin-depolymerase) and the colonization of host tissue by hyphae. Infection can occur with or without wounding of the host tissue. The entire cycle from infection to the production of new sporangiola on diseased tissue can occur in one day.

Control

Except in some tropical and subtropical areas in certain seasons, a proactive, integrated control program is not needed. In areas where Choanephora blight occurs, control is difficult during warm periods with high humidity due to frequent rain or overhead irrigation. Any tactic that provides increased aeration in the crop canopy is beneficial. Increased plant spacing, avoidance of unnecessary overhead irrigation, adequate drainage in fields, and avoidance of excess fertility, which creates a dense canopy, are useful for suppressing the disease.

Some fungicides (e.g., copper types) may provide limited control of Choanephora blight, but delivering a fungicide inside a dense canopy can be difficult. Overall, fungicide applications have not been very effective.

In gardens, removing the spent corolla after the fruit is set may be beneficial. This technique has been used in controlling Choanephora fruit rot of squash. Pepper cultivars in which the corolla abscises soon after fruit set should be less susceptible.

Selected References

Barnett, H. L., and Lilly, V. G. 1950. Influence of nutritional and environmental factors upon asexual reproduction of *Choanephora cucurbitarum* in culture. Phytopathology 40:80–89.

Benny, G. L., Humber, R. A., and Morton, J. B. 2001. Zygomycota: Zygomycetes. Pages 113–146 in: The Mycota VII. Part A, Systematics and Evolution. D. J. McLaughlin, E. G. McLaughlin, and P. A. Lemke, eds. Springer-Verlag, Berlin.

Dougherty, D. E. 1979. Bud rot of pepper. Proc. Fla. State Hortic. Soc. 92:103–106.

Sinha, S. 1940. On the characters of *Choanephora cucurbitarum* Thaxter on chillies (*Capsicum* spp.). Proc. Indian Acad. Sci. Sect. B 11: 162–166.

Thaxter, R. 1903. A New England *Choanephora:* Contributions from the cryptogamic laboratory of Harvard University. Rhodora 5:97–108.

Wolf, F. A. 1917. A squash disease caused by *Choanephora cucurbitarum.* J. Agric. Res. 7:319–327.

Yu, M. Q., and Ko, W. H. 1996. Mating type segregation in *Choanephora cucurbitarum* following sexual reproduction. Can. J. Bot. 74:919–923.

(Prepared by T. A. Kucharek,
G. L. Benny, and K. Pernezny)

Damping-Off and Root Rot

Damping-off is common in pepper in temperate and tropical climates worldwide. The disease affects seeds, emerging seedlings in transplant production, and transplants and seedlings in the field. Minor damage may occur in older plants.

Symptoms

Symptoms of damping-off of pepper vary with the stage of plant development at the time of infection. Seeds infected before germination do not germinate but turn brown and soft, become shrunken, and decompose. Seedling emergence is reduced. On infected young seedlings, brown, water-soaked lesions form about 1 cm above and below the soil line (Plate 15). The basal part of the stem is constricted and softer than the upper part and cannot support the seedling, which falls over, withers, and dies. The root system is reduced and turns brown and rots, with few or no secondary roots. In a stand of transplants, plants will be missing either in patches or at points scattered throughout the stand.

Some pathogens that cause damping-off, such as some species of *Pythium,* can infect mature plants, causing chlorosis, stunting, and reduced vigor. The roots of affected plants are brown and rotten.

Symptoms of damping-off may be confused with injury caused by excessive fertilization, high levels of soluble salts, water stress, or abiotic stresses, such as excessive heat, cold, fuel fumes, or pesticides. However, symptoms of damage due to abiotic causes are apparent on leaves before damage to roots occurs.

Causal Organisms

Damping-off and root rot are caused by several fungi, oomycetes, and bacteria, one or more of which may affect a particular crop. Most commonly reported in pepper are *Rhizoctonia* spp., including *R. solani* Kühn; *Phytophthora* spp.; *Fusarium* spp.; *Pythium* spp., including *P. ultimum* Trow, *P. myriotylum* Drechs., *P. aphanidermatum* (Edson) Fitzp., and others; and *Pseudomonas* spp.

In general, species of *Pythium* and *Phytophthora* are more likely to cause damping-off in cool, wet soils, whereas species of *Rhizoctonia* and *Fusarium* may cause the disease under warmer and drier conditions. However, there are exceptions. For example, some species of *Pythium* can cause damping-off in warmer environments. *Rhizoctonia* and *Fusarium* spp. generally cause postemergence damping-off by killing the seedling at the soil line. *Pythium* spp. attack below the soil line, often at the root tips. *P. myriotylum* and *P. aphanidermatum* cause the most root rot and growth reduction in mature plants.

Pythium species produce white, cottony colonies on culture media or in water. The mycelium is coenocytic and may be straight or sinuous. These fungi reproduce asexually by forming sporangia, which are of various sizes and shapes. Morphological characteristics of the sporangia are used in the identification of species. *P. aphanidermatum* and *P. myriotylum* produce lobate sporangia and swollen, lobed hyphae. Other species of *Pythium* produce spherical sporangia, which are either terminal or intercalary. To reproduce sexually, *Pythium* species form oogonia and antheridia, the morphology of which varies among species.

R. solani and other species of *Rhizoctonia* vary greatly in appearance. Colonies may be white but are generally varying shades of brown. The mycelium is hyaline to light brown. It is characteristically branched at right angles, frequently with a slight constriction and a dolipore septum near each branch. The fungus enters a basidial (sexual) stage on certain hosts and in culture under certain conditions.

Descriptions of other damping-off pathogens are presented in other sections of the compendium.

Disease Cycle and Epidemiology

Rhizoctonia spp. are common, endemic soilborne fungi with broad host ranges. They survive indefinitely in soils by colonizing organic material and producing sclerotia. They can be dispersed in contaminated soil or on farm equipment.

Pythium spp., like *Rhizoctonia* spp., have broad host ranges and are soilborne. They survive in the field on other hosts, including weeds. Some *Pythium* spp. produce oospores that exist in soil, independent of a host, for very long periods (more than one year). *Pythium* spp. can be spread in contaminated soil or in irrigation or ground water. Seed diffusates and root exudates stimulate the germination of oospores, with resulting infection of the host. Crowding of plants and excess nitrogen fertilization favor infection.

The disease cycles and epidemiology of diseases caused by *Fusarium* spp., *Phytophthora* spp., and *Pseudomonas* spp. are discussed in other sections of the compendium.

Control

Greenhouse soil should be treated by steam pasteurization to maintain a temperature of 71°C for 30 min. Pathogen-free soilless medium should not be allowed to become contaminated by other sources. Pots and transplant trays should be new or disinfested before reuse by hot water or steam (heating them to 71°C for 30 min) or by being placed in a 10% sodium hypochlorite solution for 30 min. Workers should clean their hands and tools before handling healthy plants. Fungicide-coated seeds with a high germination rate should not be planted too deep.

Transplant houses should maintain moderate temperatures to promote seed germination and growth. Adequate air circulation, light, and ventilation should be provided. Plants should not be overcrowded. They should have adequate water, preferably applied in the late morning.

In the field, seeds or transplants should be placed in well-drained soil with adequate spacing. Only healthy transplants should be planted. Fumigation or solarization of the soil reduces losses to damping-off. Planting should be conducted when temperatures are favorable for rapid plant growth. Movement of infested soil or infected plants should be avoided. Biocontrol agents may be beneficial in disease suppression, either as seed treatments or field applications. A fungicide may be banded in beds by mixing it with transplant water during setting. An effective fungicide should be applied as a soil drench or heavy spray as soon as the first symptoms of damping-off are observed. However, identification of the causal organism is necessary for selecting the proper fungicide. Several applications of a fungicide may be necessary, as peppers may be sensitive to some fungicides applied at a high rate.

Selected References

Agrios, G. N. 1997. Plant Pathology. 4th ed. Academic Press, San Diego, Calif.

Chellemi, D. O., Mitchell, D. J., Kannwischer-Mitchell, M. E., Rayside, P. A., and Rosskopf, E. N. 2000. *Pythium* spp. associated with bell pepper production in Florida. Plant Dis. 84:1271–1274.

Lewis, J. A., and Larkin, R. P. 1998. Formulation of the biocontrol fungus *Cladorrhinum foecundissimum* to reduce damping-off diseases caused by *Rhizoctonia solani* and *Pythium ultimum*. Biol. Control 12:182–190.

Mao, W., Lewis, J. A., Lumsden, R. D., and Hebbar, K. P. 1998. Biocontrol of selected soilborne diseases of tomato and pepper plants. Crop Prot. 17:535–542.

Waterhouse, G. M., and Waterston, J. M. 1996. *Pythium myriotylum*. Descriptions of Pathogenic Fungi and Bacteria, no. 118. Commonwealth Mycological Institute, Kew, England.

(Prepared by P. D. Roberts)

Fusarium Stem Rot

Fusarium stem rot of greenhouse peppers was first reported in Canada and England in 1994. The disease was first identified in Florida greenhouse peppers in 1999, in an outbreak in which up to 40% of the plants in the affected greenhouse range were infected. No fruit symptoms were observed in Florida, but wilting and death of the upper portions of the plant resulted in severe yield losses. The disease has not been reported as a problem in field-grown peppers.

Symptoms

In greenhouse peppers, the first symptom is a black lesion on the stem at a node where the plant has been pruned or on the stem of a plant from which fruit has been harvested. Wounds or openings in the stem provide entry sites for the pathogen. The lesion increases in length and width until it girdles the stem, causing the plant parts above it to wilt and die (Plate 16). Leaves below the lesion do not wilt.

Plants infected early in the production season have lesions at the base of the stem, while those infected later in the season are more likely to have lesions at upper nodes (Plate 17). The number of infected plants and the severity of symptoms increase throughout the growing season. Fruit damage has been reported in Canada and Great Britain but was not seen in the Florida outbreaks. However, wilted plants and branches produced no fruit, so that yields were reduced.

Causal Organism

Fusarium stem rot of peppers is caused by *Fusarium solani* (Mart.) Sacc., a fungus in the class Pyrenomycetes of the phylum Ascomycota. The vegetative stage of *F. solani* in potato–dextrose agar culture is characterized by cream-colored mycelia and crescent-shaped macroconidia. The sexual stage of the pathogen, *Nectria haematococca* Berk. & Broome, is typified by dark red, spherical perithecia (ascocarps) (Fig. 1). The peri-

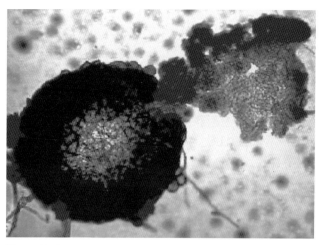

Fig. 1. Dark red, spherical perithecium (ascocarp) of *Nectria haematococca,* the sexual stage of *Fusarium solani,* the stem rot pathogen.

thecia are about 210 μm in diameter, asci are about 80 μm long, and ascospores are about 14 μm long. Mycelia and perithecia may not be obvious on diseased plants.

Disease Cycle and Epidemiology

F. solani enters plants through wounds in the stem caused by pruning and harvesting. Researchers in Canada and England reported finding mycelia and perithecia on plants and fruit, but no external evidence of the organism was noted in Florida in commercial or research settings. Very little information on the source and survival of the fungus in greenhouses has been reported. Ascospores were not found in greenhouses after the end-of-season cleanup, although they can survive for at least 14 days under typical greenhouse conditions. Perithecia in rock wool cubes in which transplants were grown were considered to be the primary inoculum. However, the source of the perithecia in the rock wool was not determined. Dissemination by ascospore discharge has been suggested.

Control

Sanitation measures offer the best control of Fusarium stem rot. Removal and disposal of infected branches or plants during the cropping season reduce the level of inoculum in the greenhouse and the spread of the disease. Using a sharp knife for pruning and harvesting is reported to reduce disease spread by promoting rapid healing of wounds. Greenhouses should be thoroughly cleaned between crops to reduce the inoculum level. The pathogen can be introduced into a greenhouse on rock wool cubes with transplants, and therefore maintaining a clean transplant production area is also important.

Controlled temperature studies show that the rate of disease development is greater at higher temperatures. High relative humidity has also been reported to encourage infection. The incidence and severity of stem rot can be reduced by environmental control using ventilation to reduce the air temperature and increase air movement through the crop.

No fungicides have been registered for control of Fusarium stem rot of pepper in the greenhouse. No differences in susceptibility to the pathogen have been observed among current cultivars.

Selected References

Copeman, R. J., and Smudja, M. 1998. Epidemiology of Fusarium stem and fruit rot of greenhouse-grown sweet pepper. On-line, ICPP Abstr. 2.4.7. Int. Congr. Plant Pathol., 7th. British Society for Plant Pathology.

Fletcher, J. T. 1994. Fusarium stem and fruit rot of sweet peppers in the glasshouse. Plant Pathol. 43:225–227.

Hanlin, R. T. 1990. *Nectria* (Fr.) Fr. Pages 132–135 in: Illustrated Genera of Ascomycetes. Vol. 1. American Phytopathological Society, St. Paul, Minn.

Jarvis, W. R., Khosla, S. K., and Barrie, S. D. 1994. Fusarium stem and fruit rot of sweet pepper in Ontario greenhouses. Can. Plant Dis. Surv. 74:131–134.

Lamb, E., Rosskopf, E., and Sonoda, R. M. 2001. First report of *Nectria haematococca* stem girdling of greenhouse peppers in Florida. Plant Dis. 85:446.

(Prepared by E. M. Lamb and E. N. Rosskopf)

Fusarium Wilt

Although there are numerous references in the literature to Fusarium wilt of *Capsicum* spp., there are very few convincing cases of a fusarial vascular wilt. The most complete documentation is by V. Rivelli, who reported the isolation of *Fusarium oxysporum* Schlechtend.: Fr. emend. W. C. Snyder & H. N. Hans. from wilted Tabasco pepper plants (*Capsicum frutescens*) grown at Avery Island, Louisiana, in 1989. The isolates recovered were pathogenic in representative accessions of several *Capsicum* spp. but were not pathogenic in cabbage, cantaloupe, cotton, cucumber, eggplant, okra, or tomato.

Other reports considered likely to be observations of a fusarial vascular wilt of pepper, in Argentina and Italy, identified *F. vasinfectum* Atk. as the causal agent. The host range of the pathogen was not reported in these studies, which preceded the reclassification of *F. vasinfectum* by Snyder and Hansen in 1940 and their designation of *F. oxysporum* f. sp. *vasinfectum* as the host-specific causal agent of Fusarium wilt of cotton and okra.

Chile blight in New Mexico was reported in 1919 to be caused by *F. annuum* Leon., but in 1960 R. M. Nakayama conclusively demonstrated that chile blight in New Mexico is caused by *Phytophthora capsici* Leon.

Various *Fusarium* spp. have been associated with root rots and cortical decays of pepper plants, but in cases in which the pathogens have been further evaluated, they have generally been shown to cause plant debilitation, but not wilting, and to have relatively wide host ranges.

Symptoms

Symptoms are described here as observed in studies of Tabasco pepper at Avery Island, where pepper plants are transplanted into the fields during April, and Fusarium wilt symptoms begin to appear about mid-June. Additional plants become infected through mid-August. The initial symptoms are a slight yellowing of the foliage and wilting of the upper leaves. The wilting progresses over the course of a few days into a permanent wilt of the plant, with the leaves still attached. The leaves of recently wilted plants are dull green but turn brown over time (Plate 18). By the time aboveground symptoms are evident, the vascular system of the plant will have become discolored with reddish brown streaks, particularly in the lower stem and roots (Plate 19). Prior to the death of the plant, the cortical tissue remains intact, with no external symptoms evident on the stem or major roots.

Causal Organism

The name of the pathogen has not been officially established, but in 1989 Rivelli proposed the name *Fusarium oxysporum* Schlechtend.: Fr. f. sp. *capsici* f. sp. nov. The cultural and morphological characteristics of the pathogen are typical of *F. oxysporum* (Fig. 2). Cultures grow and sporulate well on potato–dextrose agar at 24°C under continuous light. Colonies grown under these conditions produce aerial mycelium that is white to light tan and occasionally tinged with purple. The undersurface is peach-colored in young colonies but turns purple in older colonies. Most isolates produce macroconidia, microconidia, and chlamydospores in abundance on carnation leaf agar. Macroconidia are 36.0–52.8 × 4.1–5.0 μm and predominately three-septate. Microconidia are predominately single-celled, occasionally single-septate, and ellipsoidal to kidney-shaped. Terminal chlamydospores are produced by mycelia, and intercalary chlamydospores are produced in both mycelia and macroconidia.

Disease Cycle and Epidemiology

It is assumed that the disease cycle and epidemiology of Fusarium wilt of pepper is similar to that of diseases caused by *F. oxysporum* formae speciales that attack related crops. However, little research has been done to confirm this assumption.

The disease develops in Tabasco pepper at Avery Island during the warm summer, when average daily high and low temperatures are about 33 and 23°C, respectively. Disease development is greatly enhanced by high soil moisture. In years with near or below average rainfall during June and July (in which monthly average rainfall is 181 and 226 mm, respectively), Fusarium wilt incidence is very low. However, in years with above average rainfall during this period, 25–35% of the plants in the more poorly drained fields have been observed to wilt, in comparison to only 2–12% in better-drained fields. The plants are grown in beds 25 cm high, and most of the fields are well drained, allowing water to move away from the plants quickly after a heavy rain. Some fields have low spots, however, where the soil remains wet longer after a rain, and the incidence of Fusarium wilt is highest in these areas (Plate 20).

Fig. 2. Macroconida and microconidia of *Fusarium oxysporum*. (Reprinted, by permission, from A. R. Chase, 1987, Compendium of Ornamental Foliage Plant Diseases, American Phytopathological Society, St. Paul, Minn.)

Since the pathogen is host-specific, it is assumed that alternate hosts are not involved in its survival. The pathogen persists in crop debris, but more importantly it can survive in the soil for at least two years. The standard practice at Avery Island is to grow Tabasco pepper in a two-year rotation with another crop, usually soybean. Fusarium wilt generally occurs in a similar pattern in the same fields each time a pepper crop is grown. The pathogen can readily be isolated from field soil on a semi-selective medium; therefore, it is assumed that the pathogen can be disseminated through the movement of contaminated soil by water or equipment.

Control

Fusarium wilt is not a serious problem in many pepper production areas. Where the pathogen is present, good drainage and elevated plant beds to prevent waterlogging are effective in minimizing damage. Care should be taken to avoid moving contaminated soil on equipment or transplants from a field infested with the pathogen to an uninfested field. No resistant cultivars have been identified, but in greenhouse inoculation studies 19 *Capsicum* accessions (17 *C. baccatum,* 1 *C. chinense,* and 1 *C. annuum*) were identified as highly resistant to the Fusarium wilt pathogen.

Selected References

Curzi, M. 1927. L'eziologia della 'cancrena pedale' del *Capsicum annuum.* Riv. Patol. Veg. 17(1–2):1–18.
Jones, M. M. 1992. Fusarium wilt of pepper: Response of *Capsicum* spp. accessions to *Fusarium oxysporum* f. sp. *capsici* and analysis of vegetative compatibility. M.S. thesis, Louisiana State University, Baton Rouge.
Leonian, L. H. 1919. Fusarium wilt of chile pepper. N.M. Agric. Exp. Stn. Bull. 121.
Montemartini, L. 1907. L'avvizzimento o la malattia dei peperoni (*Capsicum annuum* L.) a Voghera. Riv. Patol. Veg. 2(17):257–259.
Nakayama, R. M. 1960. Verticillium wilt and Phytophthora blight of chile pepper. Ph.D. dissertation, Iowa State University, Ames.
Pontis, R. E. 1940. El "marchitamiento" del pimiento (*Capsicum annuum*) en la provincia de Mendoza. Rev. Argent. Agron. 7(2):113–127.
Rivelli, V. 1989. A wilt of pepper incited by *Fusarium oxysporum* f. sp. *capsici* forma specialis nova. M.S. thesis, Louisiana State University, Baton Rouge.
Snyder, W. C., and Hansen, H. N. 1940. The species concept in *Fusarium.* Am. J. Bot. 27:64–67.

(Prepared by L. L. Black)

Gray Leaf Spot

Gray leaf spot is a common disease of pepper and tomato and has a wide geographic distribution. It is most prevalent on pepper and tomato grown during the cool season in tropical and subtropical climates. In general, the disease is much more damaging to tomato than to pepper. Although gray leaf spot occurs frequently in pepper, it is often mistakenly diagnosed or overlooked, because the lesions are small and, at times, only a few lesions are present. The disease is most damaging to pepper during the seedling stage. In most cases, damage to pepper in the field is only cosmetic and does not warrant control procedures.

Symptoms

The symptoms of gray leaf spot of pepper are small lesions that form primarily on the leaf lamina but occasionally on petioles, young stems, peduncles, and calyxes. Fruit is not affected. The lesions are first evident as red to brown spots, 1–2 mm in diameter, which over time expand and turn light-colored in the center. In general, mature lesions are circular spots, typically ranging from 3 to 5 mm in diameter, with white to gray centers and red to brown margins (Plate 21). Young seedlings appear to be most susceptible, and plant beds in which transplants are grown often provide an environment conducive to disease development. Thus, it is not uncommon to observe extensive lesion development on leaves (Plate 22) and stems (Plate 23) of seedlings in plant beds. In the field, numerous lesions can form on individual leaves, causing them to turn yellow and drop (Plate 24). However, scattered lesions are often found on pepper plants throughout the season, causing no apparent injury to the crop.

Causal Organisms

Two species of *Stemphylium* have been identified as causal agents of gray leaf spot of pepper: *S. solani* G. F. Weber and *S. lycopersici* (Enjoji) W. Yamamoto (syn. *S. floridanum* Hannon & G. F. Weber).

S. solani conidiophores are pale to medium brown and up to 200 µm long and 4–7 µm in diameter, with swollen vesicles 8–11 µm in diameter. Conidia are 33–55 × 18–28 µm, pointed at the apex, muriform, pale to medium golden brown, and smooth or minutely verrucose. They have three to six transverse and several longitudinal septa and usually are constricted at the median septum. No ascigerous stage has been observed.

S. lycopersici conidiophores are pale to medium brown and up to 140 µm long and 6–7 µm in diameter, with swollen vesicles 8–10 µm in diameter. Conidia are 50–74 × 16–23 µm, conical at the apex, muriform, pale to medium brown, and smooth or minutely verrucose. They have one to eight transverse and several longitudinal septa and usually are constricted at three major transverse septa. No ascigerous stage has been observed.

S. solani was originally reported in 1930 as the causal agent of gray leaf spot of tomato and pepper in Florida. A second report from Florida, in 1955, identified *S. lycopersici* (described as *S. floridanum*) as the causal agent of a disease with similar symptoms. In both cases the pathogen studied was isolated from tomato exhibiting gray leaf spot symptoms and shown by inoculation experiments to cause gray leaf spot of pepper as well. Subsequently, natural occurrences of gray leaf spot of pepper were reported, one in Louisiana in 1958 caused by a *Stemphylium* sp. and another in Florida in 1969 caused by *S. solani.* In 1980 *S. lycopersici* isolated from tomato in the Philippines was reported to cause gray leaf spot in inoculated pepper.

Another pepper leaf spot, caused by *S. botryosum* Wallr. f. sp. *capsici* Braverman (originally reported as *S. botryosum* f. sp. *capsicum*), was described in 1968, but this disease seems to be distinct from gray leaf spot, since lesions caused by this fungus are 10–20 mm in diameter and are often composed of concentric rings.

Disease Cycle and Epidemiology

Primary inoculum can come from various sources, such as nearby fields of tomato or pepper, volunteer plants, wild solanaceous hosts, and crop debris. The pathogens are good saprophytic competitors and can persist indefinitely on plant debris. Conidia of *S. lycopersici* have been shown to survive four months at room temperature in infested crop debris and eight months in dried petri plate cultures. Optimum growth in the laboratory occurs at 26°C, and maximum sporulation at 23°C. High relative humidity favors sporulation of the pathogens, and free water is required for conidial germination. Conidia are disseminated primarily by air currents, but splashing rain can also disseminate them locally. Gray leaf spot can develop during extended periods of cloudy, wet weather, when leaves are kept sufficiently wet that conidia can germinate and penetrate the plant. The disease can also develop during clear weather if leaves are wet with dew for long periods at night.

Gray leaf spot of pepper often occurs during the seedling stage. Extended periods of cloudy, wet weather at this stage of growth can greatly increase disease severity, particularly if seedlings are grown in plant beds that hold moisture and delay drying of the foliage after watering. Under these conditions extensive defoliation can occur, which stunts the seedlings, but more critical are the deep-seated stem lesions (Plate 23). Seedlings may recover from partial defoliation after being transplanted to the field, but those with stem lesions are subject to breakage at the site of the lesions. Generally, after pepper seedlings are transplanted to the field, disease development is greatly slowed, and the plants recover fairly quickly unless they are subjected to long periods of leaf wetness from dew or rain.

Control

Most efforts to control gray leaf spot should be made at the seedling stage. Plant beds should not be established near tomato or pepper production fields, and proper care should be taken to remove crop debris and volunteer plants from the vicinity of the beds. Plant beds should also be well ventilated, to avoid the accumulation of moisture and promote rapid drying of the foliage after watering. Fungicides should be applied if disease spread cannot be checked by cultural practices such as ventilation. Once plants are in the field, it is unlikely that control measures will be required, but if they are needed, fungicide applications have been shown to be effective.

A pepper variety trial in the Philippines, conducted by the Asian Vegetable Research and Development Center in 1995, revealed that varieties vary dramatically in their response to gray leaf spot. This suggests that sources of resistance are available and can be identified if the need arises to develop resistant varieties.

Selected References

Blazquez, C. H. 1969. Occurrence of gray leafspot on peppers in Florida. Plant Dis. Rep. 53:756.

Braverman, S. W. 1968. A new leaf spot of pepper incited by *Stemphylium botryosum* f. sp. *capsicum*. Phytopathology 58:1164–1167.

Ellis, M. B., and Gibson, I. A. S. 1975. *Stemphylium lycopersici*. Descriptions of Pathogenic Fungi and Bacteria, no. 471. Commonwealth Mycological Institute, Kew, England.

Ellis, M. B., and Gibson, I. A. S. 1975. *Stemphylium solani*. Descriptions of Pathogenic Fungi and Bacteria, no. 472. Commonwealth Mycological Institute, Kew, England.

Hannon, C. I., and Weber, G. F. 1955. A leaf spot of tomato caused by *Stemphylium floridanum* sp. nov. Phytopathology 45:11–16.

Sinclair, J. B., Horn, N. L., and Tims, E. C. 1958. Unusual occurrence of certain diseases in Louisiana. Plant Dis. Rep. 42:984–985.

Valdez, R. B., and Opina, N. L. 1980. Gray leaf spot of tomato in the Philippines. Kalikasan Philipp. J. Biol. 9:81–87.

Weber, G. F. 1930. Gray leaf spot of tomato caused by *Stemphylium solani*, sp. nov. Phytopathology 20:513–518.

(Prepared by L. L. Black)

Gray Mold

Gray mold is caused by *Botrytis cinerea,* a plurivorous fungus distributed worldwide and reported to be a pathogen of plants in more than 200 genera, including *Capsicum.* The disease is common in pepper. It is generally more severe in plants grown in enclosures that maintain high relative humidity, such as plant beds or greenhouses, but it is also a threat to field-grown pepper crops. As a field disease, gray mold is most prevalent in temperate, subtropical, and highland tropical locations where cool, damp environmental conditions conducive to disease development occur most frequently. The disease is generally considered to be of minor importance in pepper, although it can affect all aboveground parts of the plant at any stage of growth and sometimes leads to substantial crop losses. Furthermore, *B. cinerea* causes an important postharvest decay of pepper (see Posthasrvest Diseases and Disorders).

Symptoms

Symptoms and signs of infection usually appear first on flower petals, at sites of injury, or on senescent tissues. Fruit symptoms begin as water-soaked spots, which rapidly expand into large, light-colored lesions (Plate 25). Characteristically, the lesions give rise to numerous long conidiophores, which are easily observed by the unaided eye. On flowers and stems of small seedlings, individual conidiophores can be seen protruding from lesions, giving the affected plant parts a spiny appearance (Plate 26). Young seedlings affected by gray mold are at risk of damping-off due to stem lesions that originate at the cotyledonary node or extend below it, or they may exhibit tip dieback as they grow a little older. Lesions on fruit, stems, and leaves of older plants are covered with thick, matted masses of gray to brown conidiophores and conidia, which are velvet-like in appearance (Plates 27 and 28).

Symptoms of postharvest decay caused by *B. cinerea* when it gains entry through mechanical wounds are similar to those on fruit attached to the plant. However, most postharvest loss attributed to *B. cinerea* is due to sunken, light-colored lesions that develop on fruit that has been chilled.

Causal Organism

Gray mold is caused by the ubiquitous fungus *Botrytis cinerea* Pers.:Fr., the conidial state of *Botryotinia fuckeliana* (de Bary) Whetzel. Conidiophores, frequently 2 mm or more in length and 16–30 μm in diameter, are pigmented near the base but hyaline in the upper portion, where they branch irregularly. One-celled conidia are produced simultaneously on numerous short denticles protruding from the apical cells of the conidiophore branches, forming a rather open head of conidial clusters. The conidia are hyaline but gray to brown in mass, ovoid, and 6–18 × 4–11 μm (mostly 8–14 × 6–9 μm). Temperatures of 18–23°C and relative humidities of 90–95% are ideal for growth, sporulation, spore release, and conidial germination. Conidia are dislodged and disseminated primarily by air currents. The fungus also produces sclerotia, which are black and flat, but their occurrence, size, and shape on natural substrata and in culture are extremely variable. Sclerotia most often germinate by producing mycelia, but they occasionally form apothecia and produce ascospores.

Disease Cycle and Epidemiology

B. cinerea overwinters as mycelium in decaying plant debris and as sclerotia in the soil, and in most areas it can persist throughout the year on some of its many hosts. It is a strong saprophytic competitor, colonizing and sporulating on senescent and dead tissues of many plant species. Conidia are mostly windborne but can also be disseminated by splashing rain. The primary inoculum appears to be conidia, in most cases, but sclerotia may also play a role by producing mycelia, which can infect plants directly. The role of ascospores, if any, in the disease cycle is not known.

Gray mold can cause much damage to seedlings in plant beds in which moisture is allowed to accumulate, as often occurs in poorly ventilated plant beds in cool, cloudy weather. Under these conditions, the pathogen can attack succulent young seedlings, causing damping-off or tip dieback. Overcrowding of plants can aggravate the situation by reducing air circulation and also by allowing the pathogen to spread by direct contact between adjacent plants. Pruning of seedlings can also increase damage due to gray mold, by creating wounds, which serve as infection sites, and clippings, which if left in the plant bed may serve as a substrate for the pathogen to colonize.

B. cinerea is a weak parasite. Gray mold rarely develops as a result of conidial infection of leaves, stems, or fruit of vigorously growing plants. The pathogen first infects and induces symptoms in weak or debilitated tissues. Therefore, after plants are established in the field or greenhouse, the focal points for initial infection and disease development are flowers, damaged tissues, and senescent leaves. Once the fungus has gained a foothold at these sites, it can move aggressively into healthy tissues and cause severe damage. For example, it can move from infected flowers or senescent leaves directly into the main stem or side branches and cause girdling stem cankers (Plate 28), which lead to top or branch dieback (Plate 29). Localized secondary spread occurs when infected plant parts, such as flowers and leaf pieces, become dislodged and fall on healthy plant organs. Contact between healthy and infected tissues allows the fungus to establish new infection sites on leaves, stems, and fruit, which conidia are unable to infect.

Postharvest losses from gray mold are intermittent and can be affected by both pre- and postharvest conditions. Cool, wet weather at or just before harvest increases the likelihood that inoculum will be present on the fruit when it is harvested. However, unless the fruit is mechanically damaged or predisposed to disease by chilling injury, gray mold is not likely to develop in storage. Postharvest cooling of pepper fruit below 13°C causes chilling injury, which predisposes the fruit to gray mold. Storage of pepper fruit at 0–4°C predisposes the fruit to the disease but suppresses disease development until the fruit is taken from storage.

Control

The control of gray mold in plant beds and greenhouses starts with sanitation. All plant debris that could serve as a substrate for the pathogen should be removed prior to sowing seed or establishing a crop. Plant beds and greenhouses should be well ventilated, to keep the humidity low. If seedlings are pruned, all clippings should be removed from the plant bed, and plant fragments should not be left attached to seedlings or lodged on them. In greenhouses, senescent leaves should be removed, and all plant debris taken out of the house. In the field, green manure should be incorporated into the soil and allowed to decompose well in advance of establishing a new crop, to minimize the amount of substrate available to *B. cinerea*.

In some cases it may be necessary to use fungicides to manage gray mold. Several fungicides have been used effectively, but fungicide-resistant strains of the pathogen have developed in some areas. Fungicides should be alternated or applied in combinations in order to reduce the risk of selecting fungicide-resistant strains.

Postharvest losses due to gray mold can be minimized by keeping storage temperatures at or above 13°C, to avoid chilling injury, which predisposes pepper fruit to decay. Hot-water treatments and ultraviolet irradiation have been reported to reduce postharvest decay caused by *B. cinerea* in bell pepper, but the practicality of these treatments is yet to be determined.

Selected References

Agrios, G. N. 1997. Plant Pathology. 4th ed. Academic Press, San Diego, Calif.

Ceponis, M. J., Cappellini, R. A., and Lightner, G. W. 1987. Disorders in fresh pepper shipments to the New York market, 1972–1984. Plant Dis. 71:380–382.

Elad, Y., Yunis, H., and Volpin, H. 1993. Effect of nutrition on susceptibility of cucumber, eggplant, and pepper crops to *Botrytis cinerea*. Can. J. Bot. 71:602–608.

Ellis, M. B. 1971. Dematiaceous Hyphomycetes. Commonwealth Mycological Institute, Kew, England.

Fallik, E., Grinberg, S., Alkalai, S., and Lurie, S. 1996. The effectiveness of postharvest hot water dipping on the control of grey and black molds in sweet red pepper (*Capsicum annuum*). Plant Pathol. 45:644–649.

Ghini, R. 1996. Occurrence of resistance to fungicides in *Botrytis cinerea* strains in the state of São Paulo. Fitopatol. Bras. 21:285–288. (CAB Abstr. 1996, Abstr. 98.)

McColloch, L. P., and Wright, W. R. 1966. Botrytis rot of bell peppers. U.S. Dep. Agric. Mark. Res. Rep. 754.

Mercier, J., Baka, M., Reddy, B., Corcuff, R., and Arul, J. 2001. Shortwave ultraviolet irradiation for control of decay caused by *Botrytis cinerea* in bell pepper: Induced resistance and germicidal effects. J. Am. Soc. Hortic. Sci. 126:128–133.

Moline, H. E. 1984. Diagnosis of postharvest diseases and disorders. Pages 17–23 in: Postharvest Pathology of Fruits and Vegetables: Postharvest Losses in Perishable Crops. H. E. Moline, ed. Northeast. Reg. Res. Publ. NE-87. Univ. Calif. Div. Agric. Sci. Bull. 1914.

Park, S. E., Lee, J. T., Chung, S. O., Kim, H. E., Park, S. H., Lee, J. T., Chung, S. O., and Kim, H. K. 1999. Forecasting the pepper grey mould rot to predict the initial infection by *Botrytis cinerea* in greenhouse conditions. Plant Pathol. J. 15:158–161.

Winstead, N. N., Wells, J. C., and Reid, W. W. 1958. *Botrytis* in pepper seedbeds and on young plants in the field. Plant Dis. Rep. 42:981–982.

(Prepared by L. L. Black)

Phytophthora Blight

Phytophthora blight is a widespread and devastating disease of pepper. The pathogen, *Phytophthora capsici,* has a wide host range among vegetable crops and can infect tomato, cucumber, melons, squash, and pumpkin. Phytophthora blight of pepper was first reported in New Mexico in 1922. Epidemics have been severe in areas of North Carolina, Florida, Georgia, Michigan, New Mexico, and New Jersey, but the disease can occur almost anywhere peppers are grown.

Symptoms

P. capsici causes a root and crown rot of pepper (Plate 30) and distinctive black lesions on stems (Plate 31). Root infection typically leads to wilting of the plant. The pathogen can also infect leaves, causing circular, grayish brown, water-soaked lesions (Plate 32). Leaf lesions and stem lesions are common when inoculum is splash-dispersed from the soil to lower portions of the plant. *P. capsici* can also infect fruit, causing lesions that are typically covered with white sporangia, a sign of the fungus (Plate 33). Nearly complete loss of plants can occur in heavily infested fields (Plate 34).

Causal Organism

Phytophthora blight is caused by the oomycete *Phytophthora capsici* Leonian. *Phytophthora* species are more closely related to nonphotosynthetic algae than true fungi and have been placed in the kingdom Chromista and the phylum Bigyra. *Phytophthora* species are water molds, and water plays a major role in the disease cycle of the pathogen.

P. capsici can be identified by both morphological and molecular methods. It can be recovered by isolation from infected tissue and plating on a semiselective medium. It can also be isolated from soil by dilution plating of soil in dilute water agar on a semiselective medium. A polymerase chain reaction (PCR) primer, PCAP, has been used with ITS 1 in PCR assays to amplify a DNA fragment of approximately 172 bp from the internal transcribed spacer region in all isolates of *P. capsici* tested from a range of hosts.

Disease Cycle and Epidemiology

P. capsici reproduces by both sexual and asexual means (Fig. 3). Oospores, the sexual spores, overwinter in the field and serve as inoculum. The disease is polycyclic within seasons. The pathogen produces two mating types, A1 and A2. They are actually compatibility types and do not correspond to dimorphic

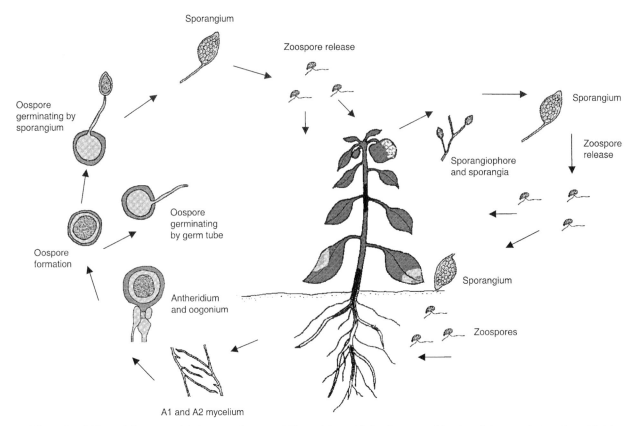

Fig. 3. Life cycle of *Phytophthora capsici,* the causal agent of Phytophthora blight of pepper. (Reprinted, by permission, from Ristaino and Johnston, 1999)

forms. Each mating type produces hormones that are responsible for gametangial differentiation in the opposite mating type. Both mating types are common in fields, and both have been identified in a single plant. *P. capsici* produces antheridia and oogonia (male and female gametangia, respectively). The antheridium is amphigynous in this species. Meiosis occurs within the gametangium, and plasmogamy and karyogamy result in the formation of oospores. Oospores germinate either directly by forming a germ tube or indirectly by producing sporangia.

The pathogen reproduces asexually by forming sporangia, which are borne on branched sporangiophores. The sporangia are generally ovoid and have a prominent papilla at the apex. They are easily dislodged from the sporangiophores and can be dispersed in fields by wind, rain, and irrigation water. Sporangia germinate indirectly and release motile, biflagellate zoospores in free moisture on plant surfaces or in saturated soil, in which the zoospores can readily move and infect roots or aboveground portions of plants.

P. capsici can infect every part of the plant. It is spread by several distinct mechanisms. Primary inoculum in the soil causes root infections, which progress to crown infections. Primary inoculum can move from root to root down a row (1) as roots grow, bringing themselves into contact with inoculum, (2) by the movement of zoospores to roots, and (3) by root-to-root contact. Inoculum can also spread down rows in surface water; this is an important means of dispersal of *Phytophthora* species causing polycyclic diseases, and it is the predominant means of dispersal of *P. capsici* in naturally infested fields. Inoculum in the soil can be spread to aboveground parts of plants by wind or by splash dispersal during rainfall or overhead irrigation. The pathogen can be rapidly spread by splashing water and in surface water, and dispersal by these means can lead to a rapid increase in disease. Inoculum is also spread by aerial dispersal from sporulating lesions on leaves, stems, and fruit.

Control

Management of Phytophthora blight currently relies on cultural practices, crop rotation, and judicious use of selective fungicides. Disease management often involves the management of soil moisture. Phytophthora blight is more severe when a high level of soil moisture is present as a result of frequent irrigation or rainfall. Appropriate site preparation, waterway systems, bed structures, and transplanting procedures that minimize water accumulation are important in Phytophthora blight management. Planting raised beds on well-drained sites can reduce the incidence of the disease. Construction of waterways that allow furrows to drain sufficiently is important. Transplanting pepper into well-formed beds and filling depressions around the transplant hole to prevent the accumulation of water can also reduce disease incidence. Subsurface drip irrigation can be more effective than more shallow drip irrigation in reducing disease in infested fields. Soil water matric potential maintained between 20 and 40 J/kg with the use of tensiometers in the field can help to avoid excessive moisture or cyclical changes in soil water potential that can stimulate the pathogen. Frequent irrigation can increase the severity of the disease.

Splash dispersal of inoculum can be reduced by planting pepper in beds mulched with straw or in the stubble of a no-till cover crop, such as rye, vetch, or wheat. If pepper is planted in beds mulched with plastic, the beds should be shaped high in the center to allow water to drain away from the plants. Zoospores can move rapidly on plastic in the rain.

Inoculum can persist in soil, so appropriate crop rotation to nonsusceptible hosts is beneficial in disease management. An infested field should not be planted with a susceptible crop for three years. Soil solarization or organic amendment of soil may reduce inoculum levels in the soil and subsequent disease.

Many pepper varieties currently grown are susceptible to Phytophthora blight. However, new varieties with resistance to the pathogen are being developed by seed companies, and some have been released. Growing a resistant variety in a management program of cultural practices and fungicide treatment can reduce the incidence of disease.

The soilborne phase of Phytophthora blight can be controlled chemically with phenylamide fungicides, such as mefenoxam. Applications of copper alone or combinations of copper-containing fungicides may provide some protection against the foliar phase of the disease. However, under climatic conditions conducive to disease development or in severe outbreaks, any fungicide is of low effectiveness at best. Resistance to the phenylamide fungicides metalaxyl and mefenoxam has been documented in some strains of *P. capsici*. New alternative classes of fungicides are needed for use in fields with a history of resistance to phenylamide compounds.

Selected References

Bowers, J. H., Sonoda, R. M., and Mitchell, D. J. 1990. Path coefficient analysis of the effect of rainfall variables on the epidemiology of Phytophthora blight of pepper caused by *Phytophthora capsici*. Phytopathology 80:1439–1446.

Parra, G., and Ristaino, J. B. 2001. Resistance to mefenoxam and metalaxyl among field isolates of *Phytophthora capsici* causing Phytophthora blight of bell pepper. Plant Dis. 85:1069–1075.

Ristaino, J. B., and Gumpertz, M. L. 2000. New frontiers in the study of dispersal and spatial analysis of epidemics caused by species in the genus *Phytophthora*. Ann. Rev. Phytopathol. 38:541–576.

Ristaino, J. B., and Johnston, S. A. 1999. Ecologically based approaches to management of Phytophthora blight on bell pepper. Plant Dis. 83:1080–1089.

Ristaino, J. B., Madritch, M., Trout, C. L., and Parra, G. 1998. PCR amplification of ribosomal DNA for species identification in the plant pathogen genus *Phytophthora*. Appl. Environ. Microbiol. 68:948–954.

(Prepared by J. B. Ristaino)

Powdery Mildew

Powdery mildew can be a serious disease of peppers in warm, arid, and semiarid growing regions. It is widely reported in Asia, Africa, the Mediterranean, and North America. The disease was first identified in bell peppers in Israel in 1950 and in chile peppers in Florida in 1971. In the 1990s, powdery mildew became a common and sometimes serious problem in California, Arizona, and New Mexico. Recent epidemics have occurred in Florida. Infection often leads to defoliation, which can lead to severe losses in pepper crops.

Symptoms

The most noticeable sign of the disease is a white, powdery growth on the underside of leaves (Plate 35). Light green to yellow lesions with necrotic centers may form on the upper leaf surface (Plate 36). Eventually the entire leaf turns pale yellow or brownish. Symptoms develop on older leaves first. When conditions are highly favorable for disease development, the pathogen may sporulate on the upper leaf surface. The edges of infected leaves eventually curl upward, revealing the fungus on the lower surface (Plate 37). Infected leaves drop prematurely from the plant, exposing fruit to the sun (Plate 38), so that they may be subject to sunscald.

Causal Organism

The causal organism, *Oidiopsis sicula* Scalia (syn. *O. taurica* E. S. Salmon; teleomorph *Leveillula taurica* (Lév.) G. Arnaud), is unique among the powdery mildew fungi. In general, powdery mildew fungi are host-specific, but *O. sicula* has an extensive host range. Powdery mildew fungi typically do not grow inside the plant but send haustoria into the plant while mycelia grow on the plant surface. In contrast, *O. sicula* produces both endophytic and epiphytic mycelia. It also has a high degree of morphological variability. This variability and the inability of some isolates to infect known hosts have led to the suggestion that *O. sicula* should be split into different species or formae speciales. At present, it remains one species, as there is no consistent way to distinguish isolates of it.

O. sicula produces hyaline mycelia, conidiophores, and single-celled conidia. Conidiophores are septate and up to 250 µm long and 8 µm wide. Conidia are cylindrical to pyriform, semi-blunt-tipped, and produced singly, in short chains, or occasionally on catenulate branches. Conidia are highly variable in size, ranging from 33 to 90 by 10 to 24 µm. Cleistothecia of the teleomorph, *L. taurica*, have been found on other host plants but have not been reported on peppers.

Disease Cycle and Epidemiology

O. sicula (*L. taurica*) has been reported to infect over 700 species in 59 plant families. With this extensive host range, the fungus can find many hosts for overseasoning and long-term survival. In addition to peppers, important crop hosts include alfalfa (*Medicago sativa*), artichoke (*Cynara scolymus*), cotton (*Gossypium hirsutum* and *G. barbadense*), eggplant (*Solanum melongena*), guar (*Cyamopsis tetragonoloba*), kenaf (*Hibiscus cannabinus*), onion (*Allium cepa*), and tomato (*Lycopersicon esculentum*). The fungus also infects ornamentals, such as mesquite (*Prosopis chilensis, P. juliflora,* and *P. glandulosa*), monkey-flower (*Mimulus aurantiacus*), and yellow bird-of-paradise bush (*Caesalpinia gilliesii*), as well as many weed species, notably annual sow-thistle (*Sonchus oleraceus*), cocklebur (*Xanthium strumarium*), desert tobacco (*Nicotiana trigonophylla*), groundsel (*Senecio vulgaris*), shepherd's-purse (*Capsella bursa-pastoris*), spurred anoda (*Anoda cristata*), white-stem filaree (*Erodium moschatum*), and Wright's ground-cherry (*Physalis wrightii*). Some cross-inoculation experiments with various isolates have been successful, but the incidence of natural cross-inoculation is unknown, although many hosts are present in major pepper-growing areas. In some locations, such as New Mexico and Arizona, many of these hosts remain uninfected with *O. sicula,* even plants growing in close proximity to infected peppers. This observation may indicate some host specialization among isolates of the fungus.

Powdery mildew of pepper occurs in both dry and humid climates. Conidia of *O. sicula* can germinate at any relative humidity (RH), from 0 and 100%, when the temperature is between 10 and 35°C. Under optimum environmental conditions (nighttime RH of 90–95%, daytime RH above 85%, and temperatures between 15 and 25°C), conidia germinate and infect a host within 24–48 hr. Mycelia grow internally, producing conidiophores and conidia through stomata. The pathogen is spread by windblown conidia. Once infection has occurred, warm days (above 30°C) with cool, humid nights (below 25°C) favor rapid disease development. The incidence of powdery mildew in pepper is greatest in humid climates, but defoliation of infected plants is more severe in dry climates. Disease incidence is greater in furrow- or drip-irrigated fields than in sprinkler-irrigated fields.

Control

Because of the wide host range of *O. sicula,* sanitation (the removal and destruction of infected crop debris and effective weed management) in and around pepper fields is not always sufficient to control powdery mildew. Resistance or tolerance varies among pepper cultivars. Many nonpungent peppers appear to have a high degree of tolerance, whereas most pungent peppers have little or no resistance to the fungus or tolerance for it. In areas where disease losses can be severe, control generally requires the use of registered fungicides. Effective chemical control depends on early detection of the disease and thorough application coverage, with the fungicide reaching the underside of the leaves and the lower plant canopy.

Selected References

Blazquez, C. H. 1976. A powdery mildew of chilli caused by *Oidiopsis* sp. Phytopathology 66:1155–1157.

Correll, J. C., Gordon, T. R., and Elliott, V. J. 1987. Host range, specificity, and biometrical measurements of *Leveillula taurica* in California. Plant Dis. 71:248–251.

Hirata, K. 1968. Notes on host range and geographical distribution of the powdery mildew fungi. Trans. Mycol. Soc. Jpn. 9:73–88.

Mihail, J. D., and Alcorn, S. M. 1984. Powdery mildew (*Leveillula taurica*) on native and cultivated plants in Arizona. Plant Dis. 68:625–626.

Nour, M. A. 1958. Studies on *Leveillula taurica* (Lev.) Arn. and other powdery mildews. Trans. Br. Mycol. Soc. 41:17–38.

Reuveni, R., and Rotem, J. 1973. Epidemics of *Leveillula taurica* on tomatoes and peppers as affected by the conditions of humidity. Phytopathol. Z. 76:153–157.

Shifriss, C., Pilowsky, M., and Zacks, J. M. 1992. Resistance to *Leveillula taurica* mildew (= *Oidiopsis taurica*) in *Capsicum annuum*. Phytoparasitica 20:279–283.

(Prepared by N. Goldberg)

Southern Blight

Southern blight is a common and destructive disease of peppers in warm, humid regions worldwide. The host range of the pathogen is extensive, encompassing at least 500 species in 100 plant families, including many vegetable, agronomic, and ornamental crops. Southern blight occurs primarily in the tropics, subtropics, and warm temperate regions in the southern United States, Central and South America, Australia, southern Europe bordering the Mediterranean, Africa, Asia, and Hawaii.

Symptoms

The pathogen attacks the stem of the pepper plant near the ground, causing the plant to turn yellow and wilt. Within days, the stem of the plant turns brown and decays both above and below the soil line. The lower part of an infected stem remains intact, and the rest of the plant wilts and turns brown. The fungus is often evident as a white mycelial mat that grows on the stem and the surrounding soil (Plate 39). Embedded in the mycelia are sclerotia, which initially appear white and fuzzy but gradually turn into smooth, hard, round objects, which are light tan to dark brown and similar in size to a cabbage seed. Pepper fruit can become infected by contact with infested soil. Lesions on fruit are water-soaked and filled with mycelia and sclerotia.

Causal Organism

Southern blight is caused by *Sclerotium rolfsii* Sacc., the sclerotial state of *Athelia rolfsii* (Curzi) Tu & Kimbrough (syns. *Corticium rolfsii* Curzi and *Pellicularia rolfsii* (Curzi) E. West). *S. rolfsii* grows rapidly in culture and within 48 hr may cover a petri plate with white mycelia containing slender aerial strands. Both in culture and in plant tissue, the mycelial mat grows outward in a fan shape. The characteristic white mycelial fans with brown sclerotia extending from infected tissues are diagnostic signs of the pathogen. At least two types of hyphae are produced. At the edge of the colony, the hyphae are large, straight cells (4.5–9 × 350 µm) with one or more clamp connections at each septum. Secondary and tertiary hyphae are slender (1.5–2.5 µm in diameter), grow irregularly, and lack clamp connections.

Sclerotia begin to form on the surface of the mycelial mat after four to seven days of growth (Plate 40). They are initially fuzzy, white mycelia, which melanize to turn dark brown when mature. The mature sclerotium (1–2 mm in diameter) has a hard rind, which protects the cortex and viable hyphae inside.

Although uncommon, a sexual fruiting stage has been reported, in which basidia are produced. Each basidium contains two or four thin-walled, colorless basidiospores, borne on short spines at the ends of slightly enlarged, short threads.

Disease Cycle and Epidemiology

S. rolfsii survives in soil for long periods in the form of sclerotia, which also serve as the primary inoculum. Sclerotia buried deep in the soil may survive for a year or less, whereas those at the surface remain viable and may germinate in response to alcohols and other volatiles released from decomposing plant material. *S. rolfsii* survives on plant debris, volunteers, and weeds. Sclerotia and mycelia are disseminated by infested soil, contaminated tools and machinery, infected transplants, irrigation water, and possibly seeds.

Environmental conditions that favor the pathogen and disease development are high temperature, high moisture, and acid soil. *S. rolfsii* rarely occurs where daily winter temperatures are below freezing. Optimal temperatures for growth and sclerotial formation are 25 to 35°C, and little or no growth occurs at or below 10°C or above 40°C. Mycelia can survive at 0°C, but sclerotia can survive temperatures at least as low as −10°C. High soil moisture favors fungal growth. Optimum mycelial growth occurs at pH 3.0 to 5.0. Sclerotia can germinate at pH 2.0–5.0, but germination is inhibited at pH 7.0 and above. Southern blight is usually not a problem in plants growing in calcareous soils with high pH.

Infection by *S. rolfsii* is usually restricted to plant parts in contact with the soil. Stems, roots, leaves, and fruit can become infected. Usually a considerable mass of mycelium is produced on the plant surface 2 to 10 days prior to infection. The pathogen penetrates host tissue by producing an enzyme that deteriorates the outer cell layer of the host tissue and causes it to decay. Further production of mycelium and the formation of sclerotia complete the cycle.

Control

Control of southern blight is difficult and depends on a variety of methods. Avoidance of the disease by selecting fields that are free of *S. rolfsii* is the most successful method of control. Diseased transplants should not be introduced into the field. Sanitation and cultural control methods may help to limit southern blight. Diseased plants should be rogued, and weeds eliminated. Plants should not be injured during cultivation. Disease incidence is greater in plantings with dense canopies, and therefore wide spacing of plants can help to lower it. Choosing a planting date timed to avoid a wet, warm period may also help to reduce disease incidence. Because of the wide host range of the pathogen, crop rotation must involve a nonhost, such as grass, corn or wheat. Soil pH can be raised by liming. Disease incidence is lower when ammonium rather than nitrate fertilizer is applied.

The fungus is highly aerobic, and burying infected plant debris and sclerotia by deep plowing (at least 20 cm deep) with a moldboard extension that inverts soil can reduce the level of inoculum. Buried soil must not be brought back to the surface during the growing season.

Black plastic mulch and row covers reduce disease severity and incidence by providing a barrier between the plants and the soil. Oat or straw compost or straw added to the soil also reduces disease incidence. The addition of these soil amendments may increase populations of antagonistic soil microorganisms.

Soil can be solarized by covering it with a transparent polyethylene sheet, which generates heat by trapping radiation in the bed, killing sclerotia in the top 1 cm of soil.

Several commercially available antagonistic fungi have been shown to reduce damage by *S. rolfsii,* including *Trichoderma harzianum, T. viride, Bacillus subtilis, Penicillium* spp., and *Gliocladium virens.* Soil solarization together with the addition of *T. harzianum* has been found to be more effective in reducing disease incidence than either treatment alone.

Preplant chemicals and fumigants, such as metam-sodium, Vorlex (a mixture of dichloropropane, 1,3-dichloropropene, and methyl isothiocyanate), methyl bromide, and chloropicrin, applied to soil, can reduce the pathogen population and control southern blight. However, untreated soil that contains *S. rolfsii,* such as soil between beds in plastic mulch fumigation systems, should not be splashed on the plants.

Selected References

Beute, M. K., and Rodríguez-Kábana, R. 1981. Effects of soil moisture, temperature, and field environment on survival of *Sclerotium rolfsii* in Alabama and North Carolina. Phytopathology 71:1293–1296.

Brown, J. E., Stevens, C., Osborn, M. C., and Bryce, H. M. 1989. Black plastic mulch and spunbonded polyester row cover as method of southern blight control in bell pepper. Plant Dis. 73:931–932.

Gurkin, R. S., and Jenkins, S. F. 1985. Influence of cultural practices, fungicides, and inoculum placement on southern blight and Rhizoctonia crown rot of carrot. Plant Dis. 69:477–481.

Jenkins, S. F., and Averre, C. W. 1986. Problems and progress in integrated control of southern blight of vegetables. Plant Dis. 70:614–619.

Mihail, J. D., and Alcorn, S. M. 1984. Effects of soil solarization on *Macrophomina phaseolina* and *Sclerotium rolfsii.* Plant Dis. 68:156–159.

Mordue, J. E. M. 1974. *Corticium rolfsii.* Descriptions of Pathogenic Fungi and Bacteria, no. 410. Commonwealth Mycological Institute, Kew, England.

Peeples, J. L., Curl, E. A., and Rodríguez-Kábana, R. 1976. Effect of the herbicide EPTC on the biocontrol activity of *Trichoderma viride* against *Sclerotium rolfsii.* Plant Dis. Rep. 60:1050–1054.

Punja, Z. K. 1985. The biology, ecology, and control of *Sclerotium rolfsii.* Ann. Rev. Phytopathol. 23:97–127.

Weber, G. F. 1931. Blight of carrots caused by *Sclerotium rolfsii,* with geographic distribution and host range of the fungus. Phytopathology 21:1129–1140.

(Prepared by P. D. Roberts)

Verticillium Wilt

Verticillium wilt is a serious disease of diverse plant species. The causal agents, *Verticillium albo-atrum* and *V. dahliae,* are ubiquitous soilborne pathogens. The disease has been reported in pepper (*Capsicum annuum*) in Europe, the Mediterranean, Canada, and the United States. It affects both field-grown and greenhouse-grown crops. The incidence and severity of the disease vary from year to year and from one location to another, depending on host susceptibility, pathogen virulence, soil type, and environmental conditions.

Symptoms

The first symptoms of Verticillium wilt of pepper are stunting and a slight yellowing of the lower foliage (Plate 41). As the disease progresses, leaves may become more severely yellowed and may drop off the plant (Plate 42). Symptom severity is highly dependent on soil and air temperatures and nutrient availability. The pathogen invades xylem elements and disrupts water transport. As the disease develops, varying degrees of vascular discoloration may occur (Plate 43), and plants wilt as a result of water stress. In some cases, plants become prematurely defoliated (Plate 44). For a few days, infected plants may recover at night, before permanent wilting sets in and the plants die.

Causal Organisms

Verticillium wilt of pepper is reported to be caused by both *Verticillium albo-atrum* Reinke & Berthold and *V. dahliae* Kleb., but *V. dahliae* is more frequently reported as the causal organism in pepper. These fungi can be isolated from plant tissue on water agar or filter paper or isolated from soil on selective media. The morphological characteristics of isolates of these fungi are somewhat variable.

V. albo-atrum and *V. dahliae* are very similar. In fact, for many years *V. dahliae* isolates were classified as microsclerotia-forming strains of *V. albo-atrum.* However, most taxonomists now consider them to be separate species.

Colonies of *V. albo-atrum* on potato–dextrose agar are initially fluffy and white to grayish. Within two to three weeks of incubation, the fungus produces brownish black resting mycelia, an aggregation of pigmented, torulose hyphal cells, 3–7 µm wide. Colonies commonly contain white sectors that lack resting mycelia. Repeated subculturing of an isolate may result in a reduction or complete suspension of the production of resting mycelia. Conidiophores are mostly erect and largely hyaline but sometimes darker at the base. They produce several whorls of two to four phialides, which are 20–50 × 1.5–3 µm. Conidia are hyaline, single-celled, and ellipsoidal to cylindrical, measuring 3.5–10.5 × 2–5 µm (Fig. 4).

Colonies of *V. dahliae* on potato–dextrose agar are initially fluffy and white to grayish, eventually turning black with the formation of numerous microsclerotia, measuring 50–200 × 15–50 µm. Colonies commonly contain white sectors lacking microsclerotia. Conidiophores are erect and hyaline and produce whorls of three or four phialides. Conidia of *V. dahliae* are hyaline, single-celled, and ellipsoidal to cylindrical, measuring 2.5–6 × 1.4–3.2 µm. The formation of microsclerotia in plant tissue as well as in culture is one of the keys to taxonomic separation of the two species.

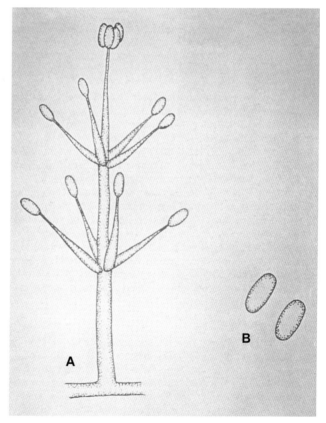

Fig. 4. Condiophore (**A**) and conidia (**B**) of *Verticillium dahliae.* (Reprinted, by permission, from T. L. Kirkpatrick and C. S. Rothrock, eds., 2001, Compendium of Cotton Diseases, 2nd ed., American Phytopathological Society, St. Paul, Minn.)

Both *V. dahliae* and *V. albo-atrum* can cause diseases of a wide range of plant species over a large geographic area. However, these fungi exist in different races or strains, which vary in virulence and host range. Most isolates of both species can infect a number of different crop plants and weeds, but a few isolates of *V. dahliae,* including the isolates from pepper, are largely host-specific or have only narrow host ranges. Most isolates of *V. dahliae* are capable of infecting chile pepper, but only isolates from bell pepper and eggplant are able to cause disease in bell pepper. Isolates of *Verticillium* from bell pepper can infect a relatively large number of hosts, while the chile pepper isolates appear to have a very restricted host range, infecting only pepper and perhaps eggplant and jimsonweed (*Datura stramonium*). The only *Verticillium* isolates that are consistently unable to infect chile pepper are isolates from cotton and cabbage, but chile pepper isolates can infect cotton. Strains isolated from the same host may vary in pathogenicity. For example, some isolates from tomato infect pepper, while others do not. Additionally, isolates from the same host may vary in pathogenicity upon reinfection of that host. While not all isolates from all geographic regions have been tested, research to date has yielded inconsistent results regarding the effect of isolate origin on host specificity and virulence. The identification of vegetative compatibility groups in *Verticillium* spp. also suggests a high degree of genetic diversity within isolates. No correlation between pathogenicity and vegetative compatibility groups has been found. Additional studies using molecular techniques to examine isolates that vary significantly in host range and pathogenicity failed to identify corresponding differences in their DNA.

Disease Cycle and Epidemiology

Verticillium survives in soil and crop debris as mycelia or microsclerotia. Microsclerotia can tolerate a wide range of environmental conditions. Most microsclerotia in soil die within two to four years, but even a small residual population is capable of causing significant crop loss. Microsclerotia have a slight ability to colonize plant debris, which may increase their population in the soil over time. Additionally, *V. dahliae* can produce microsclerotia on nonhost plants. These means of survival and population increase most likely allow the fungus to persist in soil indefinitely.

Environmental conditions that favor disease development are similar for both *Verticillium* spp., although *V. dahliae* occurs under somewhat warmer conditions (10–37°C, with an optimum temperature of 23–25°C) than *V. albo-atrum* (10–30°C, with an optimum of 21°C). Chile pepper isolates of *V. dahliae,* specifically, are favored by soil temperatures between 30 and 35°C. Both pathogens require moisture for growth and development, but *V. dahliae* appears able to tolerate dry conditions better than *V. albo-atrum.* When temperature and moisture are favorable for pathogen growth, root exudates of susceptible plants stimulate the germination of microsclerotia. The fungus then penetrates roots directly and subsequently grows through the root cortex to the xylem vessels. The xylem vessels become plugged with the fungus, so that the transport of water and nutrients is impaired, and aboveground symptoms appear as a result.

Control

At present, there are no effective measures for controlling Verticillium wilt once the disease has occurred in a field. Management strategies for avoiding the disease are most effective. Crop rotation for three or four years between pepper crops is recommended. However, the design of crop rotation is complicated by the variation in host range among isolates. Additionally, more than one isolate may be present in a field at one time. In such fields, the selection of virulent strains on other crops is a concern, though genetic changes in *Verticillium* strains appears to be slow. Regardless of the length of rotation,

some propagules of the fungus are likely to persist by saprophytic colonization of plant debris, reproduction on nonhosts, and possibly reproduction on weed species. Nevertheless, consistent or frequent pepper cropping is likely to increase the number of propagules in the soil and may lead to increased virulence within the pathogen population.

Soil fumigants containing chloropicrin have effectively controlled Verticillium wilt in many crops. Soil solarization is also effective in reducing the number of propagules in soil. Fumigation together with soil solarization may increase the rate of propagule death.

Pepper cultivars vary in their susceptibility to *Verticillium,* but no commercial cultivars resistant to the disease are currently available.

Selected References

Ben-Yephet, Y., Frank, Z. R., Malero-Vera, J. M., and DeVay, J. E. 1989. Effect of crop rotation and metam-sodium on *Verticillium dahliae.* Pages 543–555 in: Vascular Wilt Diseases of Plants. E. C. Tjamos and C. Beckman, eds. Springer-Verlag, Berlin.

Bhat, R. G., and Subbarao, K. V. 1999. Host range specificity in *Verticillium dahliae.* Phytopathology 89:1218–1225.

Domsch, K. H., Gams, W., and Anderson, T.-H. 1980. *Verticillium.* Pages 828–845 in: Compendium of Soil Fungi. Academic Press, New York.

Evans, G., and McKeen, C. D. 1975. A strain of *Verticillium dahliae* pathogenic to sweet pepper in southwestern Ontario. Can. J. Plant Sci. 55:857–859.

Green, R. J., Jr. 1980. Soil factors affecting survival of microsclerotia of *Verticillium dahliae.* Phytopathology 70:353–355.

Isaac, I. 1967. Speciation in *Verticillium.* Annu. Rev. Phytopathol. 5: 201–222.

Kendrick, J. B., Jr., and Middleton, J. T. 1959. Influence of soil temperature and of strains of the pathogen on severity of Verticillium wilt of pepper. Phytopathology 49:23–28.

Snyder, W. C., and Rudolph, B. A. 1939. Verticillium wilt of pepper, *Capsicum annuum.* Phytopathology 29:359–362.

Tjamos, E. C., Rowe, R. C., Heale, J. B., and Fravel, D. R., eds. 2000. Advances in *Verticillium* Research and Disease Management. Proc. Int. *Verticillium* Symp., 7th. American Phytopathological Society, St. Paul, Minn.

(Prepared by N. Goldberg)

White Mold

White mold, also called Sclerotinia rot, occurs sporadically in pepper. The disease also affects other crops, such as tomato, snap bean, and cabbage, in which it appears to be more serious than in pepper.

Symptoms

White mold of pepper is usually first detected at the time of flowering. The pathogen often first colonizes flower petals that have fallen into the crooks of stems and become lodged there. Water-soaked lesions subsequently appear on the stem as the pathogen invades living tissue. A soft decay of the stem follows, with zonate brown lesions, which turn whitish in younger tissue (Plate 45). Signs of the pathogen usually become apparent as the disease progresses. White, cottony mats of mycelia form on infected stems (Plate 46), and large, black sclerotia are produced within the mycelial mats. Sclerotia also form inside pepper stems (Plate 47), assuming a shape elongated in the direction of the stem cavity.

The pathogen may also initially attack plants at the soil line if senescent tissue (such as fallen leaves from plants with bacterial spot or other foliar diseases) has been deposited there. Fruit infection may occur. Water-soaked, dull green spots form on infected fruit, and cottony mycelium with embedded sclerotia forms inside the fruit.

Causal Organism

Sclerotinia sclerotiorum (Lib.) de Bary causes white mold. It is a polyphagous pathogen, attacking more than 400 plant species. It is possible that *S. minor* Jagger may also cause outbreaks of white mold, but little research in this area has been done.

Sclerotia of *S. sclerotiorum* are hard and variable in shape. Most are black outside and whitish inside and measure 3–10 × 3–7 mm. A single sclerotium may produce one or more apothecia. These sporocarps are usually cup-shaped, but some are expanded and almost flat, with a small central dimple. They are usually white, but yellow and light brown apothecia have been observed. Asci are cylindric-clavate (130 × 10 μm) and contain eight ascospores, which are nonseptate, hyaline, and ellipsoidal (9–13 × 4–5 μm). Cylindrical, hyaline paraphyses are present on the hymenium.

Disease Cycle and Epidemiology

White mold occurs in cool, damp weather. The optimum temperature for disease development is 15–21°C. High humidity and free moisture are essential for outbreaks of the disease.

Ascospores constitute virtually all primary inoculum. Sclerotia on or near the surface of the soil release airborne inoculum. The sclerotia become carpogenic after exposure to a preconditioning period of cool, moist weather. Temperatures of 11–15°C favor the formation of apothecial initials. Sclerotia can survive for many years in soil.

Ascospores are ejected from asci and wind-dispersed to susceptible pepper plants. The ascospores frequently first infest flower petals. Once the fungus becomes well established on senescent tissue, it can invade living tissue. A period of 16–72 hr of continuous shoot wetness is required for significant disease development.

Any natural or horticultural conditions that lead to poor air circulation and moisture retention in the canopy will aggravate white mold. The disease seems to be more severe in low-lying areas and in close-spaced and lush-growing crops.

Control

Application of broad-spectrum soil fumigants destroys many sclerotia of *S. sclerotiorum* in soil. However, windborne ascospores may enter pepper fields from distant points and cause considerable disease. Crop growth should be managed to avoid a dense canopy. Flooding has been used as a control measure in some fields with mineral soil in southern Florida with a history of recurring white mold. Fields must be flooded continuously for four to five weeks in the summer to significantly reduce the number of viable sclerotia. Greenhouses should be well ventilated to reduce disease incidence.

Selected References

Abawi, G. S., and Grogan, R. G. 1979. Epidemiology of diseases caused by *Sclerotinia* species. Phytopathology 69:899–904.

Knudsen, G. R., Eschen, D. J., Dandurand, L. M., and Bin, L. 1991. Potential for biocontrol of *Sclerotinia sclerotiorum* through colonization of sclerotia by *Trichoderma harzianum.* Plant Dis. 75:466–470.

Moore, W. D. 1949. Flooding as a means of destroying the sclerotia of *Sclerotinia sclerotiorum.* Phytopathology 39:920–927.

Purdy, L. H. 1979. *Sclerotinia sclerotiorum:* History, diseases and symptomatology, host range, geographic distribution, and impact. Phytopathology 69:875–880.

Yanar, Y., Sahin, F., and Miller, S. A. 1996. First report of stem and fruit rot of pepper caused by *Sclerotinia sclerotiorum* in Ohio. Plant Dis. 80:342.

(Prepared by K. Pernezny,
M. T. Momol, and C. A. Lopes)

Diseases Caused by Viruses

Virus diseases are a limiting factor in pepper production in many regions of the world. Plant viruses are obligate parasites, and they maintain extremely intimate relationships with their hosts. As pathogens, viruses are always intracellular, moving from cell to cell throughout the plant by way of plasmodesmata and vascular transport systems, most commonly the phloem.

Diseases caused by viruses in many cases are detected by characteristic symptoms, such as mosaic or mottling, deformation of leaves, or stunting. However, the type and severity of the symptoms depend on many factors, including the virus or viruses causing the symptoms, the virus strain, environmental conditions, and the age of the plant at the time of infection. Some virus infections are symptomless or result in only transient expression of symptoms. A particular virus may induce distinctly different symptoms in different cultivars of the same species. It is common for a single plant to be infected with more than one virus. Some mixed infections result in interference or masking of symptoms; some cause symptoms that are distinctly different from those caused by any of the viruses individually; and some are a synergistic interaction leading to extremely severe symptoms. Moreover, symptoms of diseases caused by viruses can be similar to symptoms of nutritional deficiencies and herbicide damage. Thus, it is difficult to visually diagnose viral diseases and identify with certainty the viruses responsible.

Management of plant viruses is particularly difficult if resistant varieties are not available, as resistant pepper varieties often are not. Part of the difficulty is that there is no chemical treatment that directly interferes with viral infection. Furthermore, most plant viruses are transmitted by insects and thus are widely dispersed in nature. The ecology of many plant viruses involves overwintering or oversummering weed species, many of which are perennials. The combination of natural plant communities harboring viruses and flying insect vectors creates a complex pathosystem. Attempts to manage some of the pepper-infecting viruses through management of their insect vectors is further complicated by the ability of insects to develop resistance to insecticides (a common problem with whitefly vectors) or to transmit viruses so rapidly that insecticides are not effective (e.g., transmission by aphids in a nonpersistent manner). These factors, taken together, along with the sporadic nature of virus outbreaks, greatly complicate management strategies that a grower can easily and predictably implement against virus pathogens.

This section describes virus diseases known to commonly occur worldwide as well as some that are documented on a regional level. Viruses known to infect pepper are listed in Table 1.

Selected References

Edwardson, J. R., and Christie, R. G. 1997. Viruses Infecting Peppers and Other Solanaceous Crops. Vols. 1 and 2. University of Florida Agricultural Experiment Station, Institute of Food and Agricultural Sciences, Gainesville.

TABLE 1. Viruses Known to Infect *Capsicum* spp.

Virus[a,b]	Genus	Other means of transmission[c]	References
Aphid-transmitted viruses, nonpersistent mode of transmission			
Alfalfa mosaic virus	*Alfamovirus*	M, S	Berkeley, G. H., 1947, Phytopathology 37:781
Broad bean wilt virus 1, Broad bean wilt virus 2	*Fabavirus*	M	Hull, R., 1969, Adv. Virus Res. 15:365
			Gracia, O., and Feldman, J. M., 1976, J. Phytopathol. 85:227
Chilli veinal mottle virus	*Potyvirus*	S	Ong, C. A., et al., 1979, MARDI Res. Bull. 7:78
Cucumber mosaic virus	*Cucumovirus*	M, S	Douine, L., et al., 1979, Ann. Phytopathol. 11:439
			Doolittle, S. P., and Zaumeyer, W. J., 1953, Phytopathology 43:333
Eggplant severe mottle virus (T)	*Potyvirus*	M	Ladipo, J. L., et al., 1988, J. Phytopathol. 122:359
Henbane mosaic virus	*Potyvirus*	M	Lovisolo, O., and Bartels, R., 1970, J. Phytopathol. 69:189
Indian pepper mottle virus (T)	*Potyvirus*	M	Sandhu, K. S., and Chohan, J. S., 1979, Indian J. Mycol. Plant Pathol. 9:177
Marigold mottle virus (T)	*Potyvirus*	M	Naqvi, Q. A., et al., 1981, Plant Dis. 65:271
Peanut stunt virus	*Cucumovirus*	M, S	Mink, G. I., et al., 1969, Phytopathology 59:1625
Pepper mild mosaic virus (T)	*Potyvirus*	M	Ladera, P., et al., 1982, J. Phytopathol. 104:97
Pepper mottle virus	*Potyvirus*	M	Nelson, M. R., and Wheeler, R. E., 1978, Phytopathology 68:979
			Purcifull, D. E., et al., 1975, Phytopathology 65:559
Pepper severe mosaic virus	*Potyvirus*	M	Feldman, J. M., and Gracia, O., 1977, J. Phytopathol. 89:146
Pepper vein banding virus (T)	*Potyvirus*	M	Dale, W. T., 1954, Ann. Appl. Biol. 41:240
Pepper veinal mottle virus	*Potyvirus*	M	Brunt, A. A., and Kenten, R. H., 1971, Ann. Appl. Biol. 69:235
Peru tomato mosaic virus	*Potyvirus*	M	Fernandez-Northcote, E. N., and Fulton, R. W., 1980, Phytopathology 70:315
Potato aucuba mosaic virus	*Potexvirus*	M[d]	Dykstra, T. P., 1939, Phytopathology 29:917
Potato virus A	*Potyvirus*	M	Fribourg, C. E., 1979, Phytopathology 69:441
Potato virus Y	*Potyvirus*	M	Ferguson, I. A. C., 1951, Plant Dis. Rep. 35:102
			Simons, J., 1956, Phytopathology 46:53
Tobacco etch virus	*Potyvirus*	M	Steepy, T. L., et al., 1967, Plant Dis. Rep. 51:709
Tomato aspermy virus	*Cucumovirus*	M, S	Brierley, P., 1955, Plant Dis. Rep. 39:152
Aphid-transmitted viruses, persistent mode of transmission			
Beet western yellows virus	*Luteovirus*	. . .	Duffus, J. E., 1981, Phytopathology 71:193
Carrot mottle virus	*Umbravirus*	M[d]	Stubbs, L. L., 1952, Aust. J. Biol. Sci. 5:399
Potato leaf roll virus	*Polerovirus*	G	Natti, J. J., et al., 1953, Am. Potato J. 30:55
			Dykstra, T. P., 1930, Phytopathology 20:853
Beetle-transmitted viruses			
Andean potato mottle virus	*Comovirus*	M	Valverde, R. A., et al., 1995, Plant Dis. 79:421
Belladonna mottle virus	*Tymovirus*	M	Lee et al., 1979, Phytopathology 69:985
Eggplant mosaic virus	*Tymovirus*	M, S	Dale, W. T., 1954, Ann. Appl. Biol. 41:240
Physalis mottle virus	*Tymovirus*	M	Lee, R. F., et al., 1979, Phytopathology 69:985
Fungus-transmitted viruses			
Indian peanut clump virus	*Pecluvirus*	M, S	Reddy, D. V. R., et al., 1983, Ann. Appl. Biol. 102:305
Potato mop-top virus	*Pomovirus*	M	Harrison, B. D., and Jones, R. A. C., 1970, Ann. Appl. Biol. 65:393
Tobacco necrosis virus	*Necrovirus*	M	Price, W. C., 1940, Am. J. Bot. 27:530
Leafhopper-transmitted virus			
Beet curly top virus	*Curtovirus*	M	Severin, H. H. P., 1929, Hilgardia 3:595
Nematode-transmitted viruses			
Pea early-browning virus	*Tobravirus*	M, S	Lockhart, B. E. L., and Fischer, H. U., 1976, Phytopathology 66:1391
Pepper ringspot virus	*Tobravirus*	M	Robinson, D. J., and Harrison, B. D., 1989, Descriptions of Plant Viruses, no. 347, Commonw. Mycol. Inst., Assoc. Appl. Biol.
Tobacco rattle virus	*Tobravirus*	M, S	Semancik, J. S., 1966, Phytopathology 56:1190

(continued on next page)

van Regenmortel, M. H. V., Fauquet, C. M., Bishop, D. H. L., Carstens, E. B., Estes, M. K., Lemon, S. M., Maniloff, J., Mayo, M. A., McGeoch, D. J., Pringle, C. R., and Wickner, R. B. 2000. Virus Taxonomy. Seventh Report of the International Committee on Taxonomy of Viruses. Academic Press, San Diego, Calif.

(Prepared by J. F. Murphy and C. E. Warren)

Alfalfa mosaic virus

Alfalfa mosaic virus (AMV) occurs worldwide and infects a wide range of crops and weeds. It causes substantial losses in pepper crops in eastern European countries, such as Bulgaria, Hungary, and Yugoslavia. Some infection occurs every year in the western United States, but the disease is usually not economically important, except in peppers grown near alfalfa fields. AMV can cause pepper yield losses of up to 65%.

Symptoms

Typical symptoms of pepper plants infected with AMV are bright yellow mosaic of leaves or white blotches in a mosaic pattern on leaves (Plate 48). Plants infected when young become stunted and may bear distorted or misshapen fruit.

Causal Agent

AMV is a member of the genus *Alfamovirus*, in the family *Bromoviridae*. The AMV genome consists of three distinct single-stranded RNA components along with a fourth RNA component, a subgenomic messenger RNA, which encodes the viral coat protein. The complete virion consists of four bacilli-

TABLE 1. (*continued*)

Virus[a,b]	Genus	Other means of transmission[c]	References
Nematode-transmitted viruses (*continued*)			
Tobacco ringspot virus	*Nepovirus*	M, S, P	Sastry, K. S., and Nayudu, M. V., 1976, Phytopathol. Mediterr. 15:60
Tomato ringspot virus	*Nepovirus*	M, S, P	Samson, R. W., and Imle, E. P., 1942, Phytopathology 32:1037
Thrips-transmitted viruses			
Capsicum chlorosis virus	*Tospovirus*	M	McMichael, L. A., et al., 2002, Australas. Plant Pathol. 31:231–239
Impatiens necrotic spot virus	*Tospovirus*	M	DeAvila, A. C., et al., 1992, J. Phytopathol. 134:133
			Hausbeck, M. K., 1992, Plant Dis. 76:795
Tobacco streak virus	*Ilarvirus*	M	Gracia, O., and Feldman, J. M., 1974, J. Phytopathol. 80:313
Tomato spotted wilt virus	*Tospovirus*	M	Cho, J. J., et al., 1989, Plant Dis. 73:375
Whitefly-transmitted viruses			
Chino del tomate virus	*Begomovirus*	. . .	Brown, J. K., and Nelson, M. R., 1988, Plant Dis. 72:866
Pepper golden mosaic virus complex (Serrano golden mosaic virus and Texas pepper virus)	*Begomovirus*	M	Brown, J. K., and Poulos, B. T., 1990, Plant Dis. 74:720
			Stenger, D. C., et al., 1990, Phytopathology 80:704
Pepper huasteco yellow vein virus (Pepper huasteco virus)	*Begomovirus*	. . .	Garzon-Tiznado, J. A., et al., 1993, Phytopathology 83:514
Pepper mild tigre virus	*Begomovirus*	. . .	Brown, J. K., and Nelson, M. R., 1989, Phytopathology 79:908
Sinaloa tomato leaf curl virus	*Begomovirus*	M	Brown, J. K., et al., 1993, Plant Dis. 77:1262
Tobacco leaf curl virus (chili leaf curl disease agent)	*Begomovirus*		Seth, M. L., and Dhanraj, K. S., 1972, J. Phytopathol. 73:365
			Osaki, T., and Inouye, T., 1981, Descriptions of Plant Viruses, no. 232, Commonw. Mycol. Inst., Assoc. Appl. Biol.
Tomato necrotic dwarf virus (U)	. . .	M	Larsen, R. C., et al., 1984, Phytopathology 74:795
Viruses transmitted by unknown means			
Asparagus virus 2	*Ilarvirus*	M, P	Uyeda, I., and Mink, G. I., 1981, Phytopathology 71:1264
Carnation Italian ringspot virus	*Tombusvirus*	M	Hollings, M., et al., 1970, Ann. Appl. Biol. 65:299
Celery latent virus (U)	*Potyvirus*	M, S	Bos, L., et al., 1978, Neth. J. Plant Pathol. 84:61
Cymbidium ringspot virus	*Tombusvirus*	M	Hollings, M., et al., 1977, Ann. Appl. Biol. 85:233
Eggplant mottled crinkle virus	*Tombusvirus*	M	Raj, S. K., et al., 1988, Plant Pathol. 37:599
Eggplant mottled dwarf virus (Pelargonium vein clearing virus, Pittosporum vein yellowing virus)	*Nucleorhabdo-virus*	M	El Maataoui, M., et al., 1985, Phytopathology 75:109
Lychnis ringspot virus	*Hordeivirus*	M, S, P	Bennett, C. W., 1959, Phytopathology 49:708
Moroccan pepper virus	*Tombusvirus*	M	Fischer, H. V., and Lockhart, B. E. L., 1977, Phytopathology 67:1352
Ourmia melon virus	*Ourmiavirus*	M, S	Lisa, V., et al., 1988, Ann. Appl. Biol. 112:291
Paprika mild mottle virus	*Tobamovirus*	M	Tobias, I., et al., 1982, Neth. J. Plant Pathol. 88:257
Pepper mild mottle virus	*Tobamovirus*	M, S	Wetter, C., et al., 1984, Phytopathology 74:405
Petunia asteroid mosaic virus	*Tombusvirus*	M	Koenig, R., and Kunze, L., 1982, J. Phytopathol. 103:361
Physalis mosaic virus (U)	*Tymovirus*	M	Peters, D., and Derks, A. F. L. M., 1974, Neth. J. Plant Pathol. 80:124
Potato virus X	*Potexvirus*	M	Steepy, T. L., et al., 1967, Plant Dis. Rep. 51:709
Ribgrass mosaic virus	*Tobamovirus*	M	Kovachevsky, I. C., 1963, J. Phytopathol. 49:127
Sunn-hemp mosaic virus	*Tobamovirus*	M	Capoor, S. P., 1962, Phytopathology 52:393
Tobacco mild green mosaic virus	*Tobamovirus*	M	Wetter, C., 1984, Phytopathology 74:1308
Tobacco mosaic satellite virus	Satellite virus	M	Valverde, R. A., et al., 1991, Phytopathology 81:99
Tobacco mosaic virus	*Tobamovirus*	M, S	Miller, P. M., and Thornberry, H. H., 1958, Phytopathology 48:665
Tomato bushy stunt virus	*Tombusvirus*	M, S	Fisher, H. U., and Lockhart, B. E. L., 1977, Phytopathology 67:1352
Tomato mosaic virus	*Tobamovirus*	M, S	Fletcher, J. T., 1963, Plant Pathol. 12:113

[a] Viruses for which the original publication to verify the susceptibility of *Capsicum* spp. could be obtained.

[b] T, tentative virus species, and U, unassigned virus species, according to the International Committee on Taxonomy of Viruses (van Regenmortel et al., 2000).

[c] G, grafting; M, mechanical transmission; P, pollen transmission; S, seed transmission. The manner of transmission listed is that of each virus in general, not necessarily in infections of *Capsicum*. If a virus is mechanically transmissible, in most cases it is also transmissible by grafting. In a few cases, e.g., *Potato leaf roll virus*, the virus is not mechanically transmissible but has been reported to be transmissible by grafting.

[d] Requiring a "helper" virus for aphid transmission.

form components, 18 nm wide and ranging from 30 to 56 nm long. AMV is serologically unrelated to any other virus, but there are significant genetic similarities between it and viruses in the genus *Ilarvirus*. Several strains of AMV have been reported.

Disease Cycle and Epidemiology

AMV is transmitted in a nonpersistent manner by at least 14 aphid species, including the green peach aphid (*Myzus persicae*), the pea aphid (*Acyrthosiphon pisum*), and the blue alfalfa aphid (*A. kondoi*). Vectors can acquire the virus after a few minutes of feeding on infected plants and can immediately transmit it to healthy plants. Aphids remain viruliferous for a short time. Transmission to peppers is a function of the transient aphid pressure more than the number of colonizing aphids.

AMV is also efficiently transmitted mechanically, by grafting, and through seed. Reports of transmission of the virus through pepper seed range from levels of 1–5% up to 69%.

AMV has been reported to infect dicotyledonous plant species in 21 families, including the Fabaceae and Solanaceae. In the western United States, symptomatic and nonsymptomatic alfalfa plants serve as reservoirs of AMV for transmission to pepper plants by aphids. Transmission of the virus through alfalfa seed increases the potential for alfalfa to serve as a source of inoculum.

Control

Insecticide sprays to control aphids are not effective in preventing AMV infection, since the nonpersistent transmission of the virus occurs too rapidly. Pepper varieties resistant to AMV are not currently available. Planting pepper crops near alfalfa fields or following alfalfa in the same field should be avoided. Virus-free seed and transplants should be used whenever possible.

Selected References

Brunt, A. A., Crabtree, K., Dallwitz, M. J., Gibbs, A. J., Watson, L., and Zurcher, E. J., eds. 1996–. Plant Viruses Online: Descriptions and Lists from the VIDE Database. Version 20 Aug. 1996. http://image.fs.uidaho.edu/vide/

Jaspars, E. M. J., and Bos, L. 1980. Alfalfa mosaic virus. Descriptions of Plant Viruses, no. 229. Commonwealth Mycological Institute and Association of Applied Biologists, Kew, England.

Šutić, D. D., Ford, R. E., and Tošić, M. T. 1999. Pages 128–129 in: Handbook of Plant Virus Diseases. CRC Press, Boca Raton, Fla.

(Prepared by R. Creamer)

Andean potato mottle virus
Pepper Strain

The pepper strain of *Andean potato mottle virus* (APMoV-P) infects several *Capsicum* species in Central America. Since 1990, it has been detected in Tabasco, Greenleaf Tabasco (*Capsicum frutescens*), and Habanero (*C. chinense*) peppers on farms in Costa Rica, Nicaragua, and Honduras. Infection by APMoV-P alone does not tend to result in severe symptoms, but mixed infection with other pepper viruses, such as *Tobacco etch virus* (TEV) or *Pepper mottle virus* (PepMoV), can cause severe symptoms and substantially lower yield. Other strains of APMoV cause diseases of potato in the Andean region of South America.

Symptoms

In Tabasco and Greenleaf Tabasco plants, APMoV-P induces mild foliar mosaic or mild yellow mottle (Plate 49). In some cultivars of *C. chinense,* it causes a yellow stipple and mottle on leaves (Plate 50). In cultivars of *C. annuum,* only a mild mottle tends to develop on infected plants, and some cultivars do not show symptoms even though the virus is present. In general, coinfection with APMoV-P and either PepMoV or TEV causes severe foliar mosaic and stunting in *C. annuum* cultivars.

Causal Agent

APMoV is a member of the genus *Comovirus,* in the family *Comoviridae.* It typically consists of three isometric particles, designated T, M and B, with genomic RNA molecules 1 and 2 occurring in particles B and M, respectively. Particle T contains no viral RNA. This virus is highly antigenic and occurs in high concentrations in host tissues. Vacuolate inclusions, characteristic of comovirus infections, are readily observed in strips of the epidermis of infected plants. The virus is readily transmitted by mechanical means. Sequence analysis of the RNA polymerase gene indicates that APMoV-P is a distinct strain of the virus. Only solanaceous plant species have been reported to be susceptible to this strain. Diagnosis of APMoV-P can be accomplished by agar or agarose double-diffusion, enzyme-linked immunosorbent assay (ELISA), or reverse transcriptase polymerase chain reaction (RT-PCR) tests.

Disease Cycle and Epidemiology

APMoV-P is transmitted by the banded cucumber beetle, *Diabrotica balteata,* a polyphagous beetle common in pepper fields throughout Central America and parts of North America. It is likely that the beetle plays an important role in the dis-

semination of APMoV-P in the field. Furthermore, since other solanaceous species have been reported to be experimental hosts of the virus, it is possible that solanaceous weed species in the field may serve as reservoirs of APMoV when *Capsicum* species are not present.

Control

Infection of peppers by APMoV-P has been reported only recently, and commercial cultivars of *Capsicum* species with resistance to the virus have not been reported. Eradication or reduction in numbers of solanaceous weed species that may harbor the virus and its insect vector could diminish the incidence of disease. APMoV-P is not transmitted through seed. In Central America, several pepper genotypes are grown as perennials and could serve as the source of infection in new plantings.

Selected References

Dufresne, D. J., Valverde, R. A., and Hobbs, H. A. 2000. Effects of co-infections of Andean potato mottle comovirus with two potyviruses in seven *Capsicum* genotypes. Rev. Mex. Fitopatol. 17:17–22.

Valverde, R. A., Black, L. L., and Dufresne, D. J. 1995. A comovirus affecting tabasco pepper in Central America. Plant Dis. 79:421–423.

Valverde, R. A., and Fulton, J. P. 1996. Comoviruses: Identification and diseases caused. Pages 17–33 in: The Plant Viruses: Polyhedral Virions and Bipartite RNA Genomes. B. D. Harrison, ed. Plenum, New York.

(Prepared by R. A. Valverde)

Beet curly top virus

Beet curly top virus (BCTV) infects a wide range of crops and weeds. This leafhopper-transmitted virus and the disease it induces occur throughout the arid and semiarid regions of the western United States and in the eastern Mediterranean. While the disease is endemic, severe losses due to curly top of pepper are sporadic, with the impact of the disease depending on cropping cycles, environmental factors, and leafhopper pressure.

Symptoms

Pepper plants infected at an early stage of development are severely stunted and form yellowed, thickened, crisp, rolled leaves (Plate 51). They may also have reduced fruit set. Plants infected at later stages of development have somewhat rolled, thickened leaves and stiff, brittle upper shoots and terminals.

Causal Agent

BCTV is the type member of the genus *Curtovirus,* in the family *Geminiviridae.* Viruses in this group are characterized by circular single-stranded DNA genomes encapsidated in twin spherical particles. BCTV exists primarily as three strains and variants of these strains. The virus is transmitted by the beet leafhopper, *Circulifer tenellus,* in the United States and by *C. opacipennis* in the Mediterranean. BCTV is restricted to the phloem and therefore is not mechanically transmissible.

Disease Cycle and Epidemiology

The beet leafhopper transmits BCTV in a persistent, circulative manner. It transmits the virus efficiently after feeding on infected plants for two days but can transmit it less efficiently after a shorter feeding time (2 to 20 min). A 4-hr latent period occurs in viruliferous leafhoppers before the virus can be transmitted, after which time they can inoculate healthy plants by feeding for as little as 1 min. Symptoms usually appear on pepper plants within one week of transmission. Leafhoppers commonly retain the ability to transmit BCTV for days to weeks.

BCTV is reported to infect plant species in 45 families, including the Asteraceae, Brassicaceae, Chenopodiaceae, Fabaceae, and Solanaceae. The virus has not been reported to be seed transmitted. Crop hosts in which natural BCTV infections have been reported include bean, cucurbits, pepper, spinach, sugar beet, and tomato. The leafhopper vector also feeds and breeds on an extensive range of plants from different families. Monocots have not been reported to be hosts of either the virus or the vector.

BCTV epidemiology in the western United States is determined by climate, plant diversity and distribution, and cropping cycles. The long-distance spread of BCTV by the beet leafhopper is well documented. In central California, beet leafhoppers migrate to the foothills in the late fall and overwinter on weeds. As these plants dry in the spring, the leafhoppers migrate into the valley to feed on crops and weeds, infecting them with BCTV. The beet leafhopper is thought to breed on weeds in southern New Mexico and Arizona in the spring and early summer, migrate north and east into cropping areas, and return to the breeding grounds in the late fall. In Idaho, leafhoppers can overwinter on weeds near cultivated areas, as well as migrate from desert areas to crop fields in the spring.

Control

Insecticide sprays have been used in California to reduce leafhopper numbers, but they do not prevent the transmission of BCTV. Since young plants are more severely affected by the virus, planting early or late to avoid peak leafhopper flights may help reduce losses. Removal of known weed hosts from crop fields may help to control the disease. Another disease management practice is heavy seeding of directly planted pepper, because healthy plants can compensate somewhat for the loss of infected plants. Heavy seeding also allows early removal of symptomatic plants by hand thinning. No resistance to BCTV has been identified in commercially grown peppers.

Selected References

Bennett, C. W. 1971. The Curly Top Disease of Sugarbeet and Other Plants. Monogr. 7. American Phytopathological Society, St. Paul, Minn.

Klein, M. 1992. Role of *Circulifer/Neoaliturus* in the transmission of plant pathogens. Pages 152–193 in: Advances in Disease Vector Research. Vol. 9. K. F. Harris, ed. Springer-Verlag, New York.

Stenger, D. C., and McMahon, C. L. 1997. Genotypic diversity of beet curly top virus populations in the western United States. Phytopathology 87:737–744.

(Prepared by R. Creamer)

Capsicum chlorosis virus

Capsicum chlorosis disease was first detected in pepper plants in Queensland, Australia, in 1999. The symptoms are similar to those caused by *Tomato spotted wilt virus* (TSWV) in *Capsicum* spp. but are caused by a distinct species in the genus *Tospovirus,* for which the name "Capsicum chlorosis virus" (CaCV) has been proposed. The distribution and incidence of the disease have steadily increased in all subtropical and tropical production areas of Queensland.

Symptoms

The natural hosts of CaCV identified thus far include pepper and tomato. In pepper, the symptoms resemble those caused by TSWV, but with several distinct features. Marginal and interveinal chlorosis develop on young leaves of plants infected with CaCV. These leaves often become narrow and curled, with a straplike appearance. Older leaves become chlorotic, and ring spots and line patterns may develop, like those typical of TSWV infection. Newly developed leaves of plants affected by CaCV seldom display the bright chlorotic blotches and ring spots typical of TSWV infection in *Capsicum* spp. Infection during the first four weeks of growth results in severe stunting. The fruit of infected plants is small, distorted, and frequently marked with dark, necrotic lesions and scarring over the surface. Symptomatic fruit is generally unmarketable.

Symptoms of Capsicum chlorosis disease of tomato are similar to those caused by TSWV: affected plants are stunted, and necrotic flecks and spots develop on leaves and petioles. In some cases, leaves become chlorotic and mottled, and purple ring spots form on them.

Causal Agent

Capsicum chlorosis disease is caused by a previously undescribed species in the genus *Tospovirus,* family *Bunyaviridae.* The virus reacts with antibodies to serogroup IV tospoviruses but not with antibodies to serogroups I (TSWV), II, or III. Sequence analysis of the putative N gene of the virus indicates that it is a member of serogroup IV. With less than 90% sequence identity with recognized members of the genus, it is likely to be an undescribed species. The N gene of the virus has less than 30% identity at the amino acid level with that of TSWV.

Disease Cycle and Epidemiology

Insect trapping data suggest that one or more thrips species, including *Thrips palmi* (melon thrips), *Frankliniella occidentalis* (western flower thrips), and *F. schultzei* (tomato thrips), are likely to be vectors of CaCV, but the identity of the species transmitting the virus has not been determined.

The known natural hosts of CaCV are pepper and tomato. The virus survives in infected pepper crops, ratooned crops, and out-of-season plantings in subtropical and tropical areas. Weed species that are hosts of CaCV have not yet been identified.

Disease incidence varies considerably between seasons, regions, and times of the year. Incidence has reached 50% in several instances, but levels of 5 to 10% are more common.

Control

All commercially available pepper cultivars tested are susceptible to CaCV, including those with the *Tsw* gene for TSWV resistance. Control measures are similar to those recommended for management of TSWV in *Capsicum* spp. Virus-free transplants should be planted. Old, infected crops should be destroyed, and new crops should not be placed near advanced crops with capsicum chlorosis disease. Weed control to reduce vector populations and possible virus sources is also an effective management practice.

Selected Reference

McMichael, L. A., Persley, D. M., and Thomas, J. E. 2002. A new tospovirus serogroup IV species infecting capsicum and tomato in Queensland, Australia. Australas. Plant Pathol. 31:231–239.

(Prepared by D. M. Persley)

Chilli veinal mottle virus

Chilli veinal mottle virus (ChiVMV) is an important plant virus first reported affecting chili pepper plants in Malaysia in 1979 (the spelling *chilli* is that of C. A. Ong and co-workers, who first described the virus). Since its discovery, the virus has been found in Bangladesh, Bhutan, China, India, Indonesia, Korea, Nepal, Pakistan, Sri Lanka, Taiwan, Thailand, and the Philippines. Infection of young plants can result in yield reductions of more than 50%. The average incidence of ChiVMV in farmers' fields is 20%, second only to that of *Cucumber mosaic virus,* which with 24% incidence is the most common virus affecting chili crops in Asia.

Symptoms

Symptoms induced by ChiVMV vary in type and intensity, depending on the virus strain, cultivar, and time of infection (growth stage and season). Foliar mosaic and mottle develop in early stages of systemic infection, while dark green veinbanding and leaf distortion are common during later stages of infection (Plate 52). Necrotic ring spots may appear on the leaves. Pepper plants that become infected early in their development are often stunted, their flowers drop off, and their fruit may be reduced in number and size. Additional symptoms on the fruit of ChiVMV-infected plants include a mottled appearance, various degrees of external and internal distortion, and in some cases necrosis.

Causal Agent

ChiVMV is a member of the genus *Potyvirus,* in the family *Potyviridae.* Particles of the virus are flexuous, filamentous rods, with a normal size of 793×11 nm (measured from crude sap preparations). The virus can readily be mechanically transmitted. In double-antibody sandwich enzyme-linked immunosorbent assay (DAS-ELISA), antiserum to ChiVMV reacts strongly only with its homologous antigen, whereas only weak reactions occur with *Pepper veinal mottle virus* (PVMV), *Potato virus Y,* and *Pepper mottle virus.* ChiVMV induces cytoplasmic cylindrical inclusions of type IV, indistinguishable from those induced by PVMV. Nuclear inclusions (fibrillar bundles) are only rarely observed. Small, elongated crystalline cytoplasmic inclusions are sometimes associated with ChiVMV infections. Strains of ChiVMV have been classified by the reactions of differential *Capsicum* spp.

Disease Cycle and Epidemiology

Seven species of aphids—*Aphis craccivora, A. gossypii, A. spiraecola, Myzus persicae, Toxoptera citricida, Hysteroneura setariae,* and *Rhopalosiphum maidis*—transmit ChiVMV in a nonpersistent manner. The incidence of the virus in farmers' fields does not seem to be associated with infection of weed hosts or other solanaceous crops that might serve as virus reservoirs. ChiVMV has not been found on other economically important crops, such as tomato or eggplant, which can be infected experimentally. Its host range is confined to solanaceous plants, including eggplant, petunia, *Physalis,* tobacco, and tomato. Datura, *Chenopodium amaranticolor,* and *C. quinoa* are not considered hosts of ChiVMV. Infected volunteer pepper plants in farmers' fields and kitchen gardens probably contribute to the year-round presence of the virus and serve as sources of primary inoculum. ChiVMV is not transmitted through seed.

Control

The only economical and effective control of ChiVMV is the use of resistant or tolerant cultivars. The Asian Vegetable Research and Development Center, in Taiwan, and the Malaysian Agriculture Research Development Institute have identified germ plasm resistant to one or several known strains of ChiVMV. This material has potential either for direct use by farmers or for use in breeding programs. Seedbeds are covered with a 64-mesh net to produce virus-free transplants, and thus virus infection in the field can be delayed.

Many farmers use reflective silver-coated polyethylene sheets as mulch, with weekly or more frequent insecticide applications. However, aphid populations and virus incidence have usually not been significantly reduced by this approach, nor have yields been not significantly increased.

Selected References

Asian Vegetable Research and Development Center. 2000. AVRDC Report 1999. AVRDC, Shanhua, Tainan, Taiwan.

Green, S. K. 1995. Viruses of pepper in Asia and the Pacific region. Pages 98–129 in: Proc. Int. Conf. Chilli Pepper Prod. Tropics. B. H. Chew, W. H. Loke, R. Melor, and A. R. Syed, eds. Malaysian Agricultural Research and Development Institute (MARDI), Kuala Lumpur.

Green, S. K., and Chou, J. C. 1991. Studies on management practices to reduce aphid transmitted viruses and their vectors in pepper (*Capsicum annuum* L.). Plant Prot. Bull. (Taichung) 33:364–375.

Green, S. K., Hiskias, Y., Lesemann, D.-E., and Vetten, H. J. 1999. Characterisation of chilli veinal mottle virus as a potyvirus distinct from pepper veinal mottle virus. Petria 9(3):332.

Ong, C. A., Varghese, G., and Ting, W. P. 1979. Aetiological investigations on a veinal mottle virus of chili (*Capsicum annuum* L.) newly recorded from peninsular Malaysia. MARDI Res. Bull. 7:78–88.

(Prepared by S. K. Green)

Chino del tomate virus

Chino del tomate virus (CdTV) is a whitefly-transmitted geminivirus that is widely distributed in pepper- and tomato-producing areas of Mexico. *Chino del tomate* (CdT) disease was first noticed on cultivated tomatoes grown in the coastal production areas of Sinaloa, Mexico, in 1970 and 1971. Epidemics of CdT occurred routinely in pepper and tomato crops during the 1970s and 1980s, associated with high populations of the whitefly *Bemisia tabaci,* the vector of the virus. The increase in disease incidence is thought to have been due to an expansion of vegetable production in much of Mexico and to the wide distribution of the introduced biotype B of *B. tabaci,* which prefers pepper and tomato to cucurbits and legumes more than the local biotype. The disease now occurs occasionally in most pepper and tomato fields, usually at lower incidence. CdTV infection was recently reported as far north as Nogales, Mexico, in hydroponic greenhouses.

The effect of the virus on losses in pepper production is not clear, since mixed infections by CdTV and other whitefly-transmitted and non-whitefly-transmitted viruses are common. For the same reason, CdT can be confused with other viral diseases.

Symptoms

CdTV causes mild symptoms in pepper, severe stunting and leaf curl in tomato, severe symptoms in *Datura stramonium,* and mild symptoms in bean. It causes the most severe foliar symptoms and the greatest degree of stunting in tomato of any whitefly-transmitted New World geminivirus described to date.

Under natural conditions, CdTV infection of pepper foliar tissue is nearly symptomless, but it can cause striping of bell pepper fruit. In experimentally inoculated pepper, plants exhibit mild leaf curling, mild stunting, transient yellow vein etching of leaves at outset, and a mild mosaic under high light conditions. Leaves of mature plants have distorted veins, making them slightly twisted and misshapen (Plate 53). Mild striping develops on the fruit of sweet pepper and some hot pepper varieties.

Foliar symptoms reported to be caused by a strain of CdTV designated Tomato leaf crumple virus (TLCrV) are comparable in most hosts examined, except that in tomato (at least in certain cultivars) it causes less severe curling and shortening of internodes.

Causal Agent

Partial characterization of geminate virus particles recovered from tomato plants with CdT indicated the presence of a geminivirus, which has been designated *Chino del tomate virus,* a member of the genus *Begomovirus,* in the family *Geminiviridae.* It has a bipartite genome (consisting of two genomic components), which is characteristic of other New World and some Old World begomoviruses. The two viral components, A and B, collectively consist of approximately 5.2 kb and contain six or seven genes, each of which encodes a protein. The virus uses these proteins in replication, cell-to-cell movement, systemic infection, and encapsidation of virus particles.

Complete nucleotide sequence analyses are available for CdTV and TLCrV. The latter was recently isolated from tomato in Sinaloa, Mexico, and sequence comparisons, host range, and symptomatology in common hosts indicate that the two viruses are strains of the same species.

The occurrence of other begomoviruses infecting tomatoes in the Sonoran Desert had led to speculation that *Pepper huasteco yellow vein virus* (PHYVV) together with the A component of Texas pepper virus (TPV) might cause CdT disease, but molecular characterization of the virus and the completion of Koch's postulates by inoculation of tomato plants with infectious clones demonstrated that the causal agent is a unique bipartite begomovirus, distinct from TPV and PHYVV.

CdTV and TLCrV are experimentally and naturally transmissible only by the whitefly *B. tabaci.* The virus-vector relationships are consistent with those of other begomoviruses: the vector transmits the virus in a persistent manner, after the virus has circulated through the vector to reach the salivary glands. For CdTV, the process may take 6–12 hr, after which the vector can continue to transmit the virus for the rest of its life. CdTV is not mechanically transmissible from pepper or tomato to pepper, tobacco, or tomato plants. However, TLCrV is sap-transmissible to *Nicotiana benthamiana.* CdTV and related strains are not transmissible through true seed.

CdTV naturally infects pepper, tobacco, tomato, and *Malva parviflora.* Experimental hosts include *Datura stramonium* and *Phaseolus vulgaris,* as well as pepper, tobacco, and tomato.

Disease Cycle and Epidemiology

Little is known about the epidemiology of CdTV except that it has been present in pepper and tomato plantings in Mexico since it was first noted there over 30 years ago. In nature, *M. parviflora* may serve as a host of the virus, but it is just as likely that pepper or tomato plantings are the main source of CdTV infecting crops and also *M. parviflora* when it grows near an infected crop. No other weed hosts of the virus have been identified. CdTV-infected tobacco may also be an important primary host from which virus may spread to pepper crops. This may suggest that annual infections are initiated in pepper, tobacco, or tomato, which are grown nearly continuously in this region. Because these crops are generally grown year-round and in close proximity to one another, they are thought to serve as the reservoirs for subsequent infection of cultivated species.

When the efficiency of the local biotype of *B. tabaci* (AZ A like) in transmitting CdTV was compared with that of the introduced Old World biotype B, the indigenous whitefly was the better vector.

Control

Mixtures of begomoviruses and viruses with RNA genomes are common, making it difficult to determine the importance of infection of pepper by CdTV. The virus causes less damage in pepper than in tobacco or tomato, but losses of certain types of pepper, including sweet peppers, can be substantial if plants are infected at an early growth stage. Reduction or elimination of the whitefly population in pepper crops and adjacent crops in which whiteflies breed is essential. The use of clean transplants is helpful, but once infection occurs, losses are likely to result. Removing plants that become infected early can reduce the level of inoculum in pepper fields, when tomato is not planted nearby. Weed control near fields and removal of an infected crop prior to planting a subsequent crop are advised. No resistant pepper varieties are available.

Selected References

Brown, J. K. 2001. Molecular markers for the identification and global tracking of whitefly vector–begomovirus complexes. Virus Res. 71:233–260.

Brown, J. K., and Hine, R. B. 1984. Geminate particles associated with the leaf curl or 'chino' disease of tomatoes in coastal areas of western Mexico. (Abstr.) Phytopathology 76:844.

Brown, J. K., Idris, A. M., Torres-Jerez, I., Banks, G. K., and Wyatt, S. D. 2001. The core region of the coat protein gene is highly useful for establishing the provisional identification and classification of begomoviruses. Arch. Virol. 146:1–18.

Brown, J. K., and Nelson, M. R. 1988. Transmission, host range, and virus-vector relationships of chino del tomate virus, a whitefly-transmitted geminivirus from Sinaloa, Mexico. Plant Dis. 72:866–869.

Brown, J. K., and Nelson, M. R. 1989. Two whitefly-transmitted geminiviruses isolated from pepper affected with tigré disease. (Abstr.) Phytopathology 79:908.

Brown, J. K., Ostrow, K. M., Idris, A. M., and Stenger, D. C. 2000. *Chino del tomate virus:* Relationships to other begomoviruses and identification of A-component variants that affect symptom expression. Phytopathology 90:546–552.

Gallegos, H. M. L. 1978. Enchinamiento del tomate. Page 119 in: Enfermedades de cultivos en el estado do Sinaloa. SARH, Sinaloa, Mexico.

Idris, A. M., Lee, S. H., and Brown, J. K. 1999. First report of chino del tomate and pepper huasteco geminiviruses in greenhouse-grown tomato in Sonora, Mexico. Plant Dis. 83:396.

Idris, A. M., Smith, S. E., and Brown, J. K. 2001. Ingestion, transmission, and persistence of *Chino del tomate virus* (CdTV), a New World begomovirus, by Old and New World biotypes of the whitefly vector *Bemisia tabaci.* Ann. Appl. Biol. 139:145–154.

Paplomatas, E. J., Patel, V. P., Hou, Y.-M., Noueiry, A. O., and Gilbertson, R. L. 1994. Molecular characterization of a new sap-transmissible bipartite genome geminivirus infecting tomatoes in Mexico. Phytopathology 84:1215–1224.

(Prepared by J. K. Brown)

Cucumber mosaic virus

Cucumber mosaic virus (CMV) can cause severe losses in most types of vegetable crops, including celery, cowpea, cucurbits, lettuce, pepper, and tomato, and also affects legumes and ornamentals. The virus is distributed worldwide, with a host range of approximately 1,200 plant species in 101 families, including mono- and dicotyledonous plants and many weeds, and it can be transmitted by more than 75 species of aphids. The broad host range of the virus and the large number of vector species make CMV a serious threat to plant health and plant-related economies.

Symptoms

Symptoms induced by CMV in pepper plants vary with the viral strain or isolate, the pepper genotype, and environmental conditions under which the plants are grown. Typical responses of pepper plants to CMV infection are (1) mild or no symptoms on inoculated leaves followed by mosaic on uninoculated leaves and (2) necrotic symptoms on inoculated and uninoculated leaves.

In pepper plants expressing the first type of symptoms, mechanical (rub) inoculation of older leaves with CMV under greenhouse or field conditions may not result in any apparent symptoms on inoculated leaves, although faint, diffuse chlorotic lesions may develop and then may fade with time. Generally, the diffuse chlorotic lesions are more likely to develop on plants in greenhouses. With time, under both greenhouse and field conditions, older leaves that have been mechanically inoculated with CMV become chlorotic and abscise prematurely. Inoculation of young leaves typically does not result in any visible symptoms. Systemic veinclearing of uninoculated leaves of infected pepper plants sometimes develops in greenhouses but typically is not observed in the field. Another systemic symptom often observed in pepper plants grown in greenhouses is chlorosis starting at the base of young leaves and progressing over the entire length of the leaf in a yellowish mosaic. In some pepper genotypes infected by certain strains of CMV, a more subtle but characteristic mosaic consisting of light green and dark green areas may develop. The veinclearing, chlorosis, and mosaic are fairly short-lived symptoms and are likely to develop under conditions that favor lush, succulent growth, in contrast to the more weathered appearance of leaves of plants grown in the field. Regardless of the type, extent, or severity of mosaic symptoms, they tend to fade as the leaves turn a dull green and become brittle and leathery.

In some CMV infections, small necrotic specks or ring spots with oak leaf patterns develop on inoculated leaves and sometimes on uninoculated leaves (Plate 54). The symptoms develop on both old and young leaves, and they occur whether CMV is introduced by aphids or by mechanical inoculation. On uninoculated leaves of pepper plants that are in a sink-to-source transition at the time of infection, a necrotic line often develops, running across the leaf and marking the boundary between infected tissue at the base of the leaf and uninfected tissue at its tip (Plate 55). The infected tissue at the base of the leaf typically becomes chlorotic, while the uninfected portion remains dark green. Patches of necrosis often develop on young uninoculated leaves of the same plant, and these leaves eventually turn dull green and acquire a brittle or leathery texture.

Young systemically infected leaves of pepper plants expressing either of the phenotypes described above become narrow and cease to expand (Plate 56), remaining much smaller than comparable leaves of healthy plants. Infected plants may produce the same number of leaves as healthy plants but have shortened internodes and are generally stunted. Plant top weight is substantially reduced in infected plants.

Infected plants may also produce the same number of fruit as healthy plants, but fruit size and the number of marketable fruit may be greatly reduced. Fruit symptoms include a wrinkled, bumpy appearance, a pale to yellowish green color, irregular ripening, and in some cases sunken lesions with necrotic centers (Plate 55). Necrotic lines or ring spots form on the fruit of some pepper varieties.

No extensive evaluation of interactions of CMV and satellite RNA in pepper has been reported. However, as in many other plant species, the response in pepper varies with the CMV strain, satellite RNA, and genotype of the plant. Satellite RNA has no apparent effect on CMV-induced symptoms in some cases and exacerbates symptoms in other cases.

The extent and severity of CMV infection of pepper is significantly affected by plant age at the time of infection. Plants infected while they are young typically develop severe symptoms, including mosaic or necrosis; small, deformed leaves; and severe stunting of the plants. However, as they grow older, pepper plants tend to express mature plant resistance to CMV and, as a result, may remain symptomless, with no virus detected in uninoculated leaves, or have only mild symptoms with correspondingly low levels of CMV accumulation in uninoculated leaves. These plants are also not affected by leaf deformation or stunting. Furthermore, pepper plants infected by CMV during more mature stages of growth may have few or no symptoms on fruit.

Causal Agent

CMV is the type species of the genus *Cucumovirus,* in the family *Bromoviridae.* It consists of three icosahedral particles, each approximately 28 nm in diameter. The CMV genome consists of three single-stranded, messenger-sense RNA molecules, designated RNA 1 (containing about 3,350 nucleotides), RNA 2 (about 3,050 nucleotides), and RNA 3 (about 2,200 nucleotides). Each RNA molecule is encapsidated separately as a single particle, with the RNA 3 particle containing a fourth RNA molecule, RNA 4, which is generated as subgenomic RNA from approximately 1,030 nucleotides at the 3′ terminus of RNA 3. RNAs 1 and 2 encode replication-associated proteins, 1a and 2a, respectively. RNA 2 also encodes a 3′-terminal protein, 2b, which is translated from subgenomic RNA 4A. In some hosts, 2b supports systemic infection and functions as a suppressor of gene silencing. RNA 3 encodes a movement protein, 3a, and the coat protein, which is translated from subgenomic RNA 4. In addition to structural functions, the CMV coat protein is the determinant for aphid transmission and is involved in virus movement.

CMV strains have been divided into two subgroups, designated I and II, distinguished by differences in genome sequence and serological properties. Subgroup I strains may be further divided into IA and IB, differentiated in cowpea (*Vigna unguiculata*) by inducing a systemic mosaic or necrotic local lesions on inoculated leaves, respectively.

In addition to the CMV-related RNAs, some strains of CMV support a satellite RNA, an unrelated molecule containing 335 to 405 nucleotides. The satellite RNA is dependent on CMV for encapsidation (and therefore movement from plant to plant) and replication, while serving no apparent function for the parent virus. Coinfection by CMV and satellite RNA may have no effect on symptom expression, or symptoms may be attenuated or enhanced, depending on the helper strain of CMV, the satellite, and the host. An attenuation of symptoms due to the presence of satellite RNA corresponds to a decrease in CMV accumulation. In pepper, symptoms of coinfection by CMV and satellite RNA do not appear different from symptoms caused by CMV alone, whereas in tomato the coinfection intensifies the symptoms.

Disease Cycle and Epidemiology

The broad natural host range of CMV, including many annual and perennial weed species, and the large number of aphid species that can transmit it enable the virus to remain in agroecosystems for years. *Myzus persicae* and *Aphis gossypii* are two of the more efficient vectors of CMV. However, the ubiquitous nature of the virus and the nonpersistent mode of transmission by its vectors increase the likelihood that other aphid species may introduce the virus into a crop, in which it is subsequently spread by more efficient vectors. CMV is transmitted through the seed of some weed species, in which it may be present early in the season for transmission to plants of economic importance. The introduction of the virus into crops early in the season, when plants are young and more severely affected, is an important aspect of the disease cycle and leads to severe losses of marketable fruit. While many weed species are susceptible to CMV infection, not all express symptoms or serve as good sources of inoculum. There are no reports of transmission of CMV through pepper seed.

Control

Sources of CMV resistance and tolerance have been described and, in a few cases, introgressed into commercially acceptable varieties. One example of importance is *Capsicum annuum* 'Perennial.' However, resistance tends to be multigenic and provides protection against only a small number of strains or isolates of the virus.

In the absence of resistant varieties, cultural practices that delay the introduction of the virus into a pepper crop, to exploit

Color Plates

1. Bacterial canker leaf lesions.

2. Bacterial spot leaf lesions.

3. Severe defoliation of the lower canopy of plants with bacterial spot.

4. Bacterial spot fruit lesions.

5. Bacterial wilt symptoms in the field.

6. Dark brown discoloration of vascular tissues in the lower stem of a wilted plant infected with *Ralstonia solanacearum*.

7. White, milky ooze of bacterial cells in clear water from the stem of a plant with bacterial wilt.

8. Leaf spots caused by *Pseudomonas syringae* pv. *syringae* on transplants.

9. Necrotic lesions caused by *Pseudomonas syringae* pv. *syringae* on mature leaves.

10. Anthracnose lesion in an early stage of development on a fruit.

11. Mature anthracnose lesion on a fruit, containing acervuli and spores of a *Colletotrichum* sp.

12. Anthracnose lesions on fruit.

13. Frogeye leaf spot, caused by *Cercospora capsici*.

14. Wilted plant with Choanephora blight (wet rot), caused by *Choanephora cucurbitarum*.

15. Wilted seedlings with stem lesions caused by damping-off fungi. (Courtesy R. Johnson)

16. Wilted plant with Fusarium stem rot in a greenhouse.

17. Fusarium stem rot lesion on an upper node of a stem.

18. Fusarium wilt of Tabasco pepper in an early stage of infection (right) and in an advanced stage (left).

19. Vascular discoloration in the roots and stem of a Tabasco pepper plant with Fusarium wilt.

20. Low area of a Tabasco pepper field with a high incidence of Fusarium wilt.

21. Typical leaf lesions on a field-grown plant with gray leaf spot.

22. Gray leaf spot lesions on a leaf of a seedling grown under high-moisture conditions in a plant bed.

23. Gray leaf spot lesions on the stem of a transplant in a plant bed, infected at an early stage of growth.

24. Lesions on a yellowed leaf of a field-grown plant with gray leaf spot.

25. Young lesion caused by *Botrytis cinerea* on a fruit.

26. Seedlings infected with *Botrytis cinerea,* with conidiophores on the diseased stems.

27. Gray mold fruit lesion, with masses of conidiophores and conidia of *Botrytis cinerea.*

28. Stem-girdling canker caused by *Botrytis cinerea.*

29. Dieback of branches as a result of stem cankers caused by *Botrytis cinerea.*

31. Black stem lesion caused by *Phytophthora capsici.*

30. Root and crown rot of bell pepper caused by *Phytophthora capsici.*

32. Grayish brown water-soaked leaf lesions caused by *Phytophthora capsici.*

33. Fruit lesion caused by *Phytophthora capsici,* with white mycelium and sporangia of the pathogen.

34. Plant loss in a field heavily infested with *Phytophthora capsici.*

35. Powdery mildew fungus *Oidiopsis sicula* on the underside of a leaf.

36. Early symptoms of powdery mildew on the upper surface of a leaf.

37. Leaf curl symptom of powdery mildew, caused by *Oidiopsis sicula.*

38. Premature leaf drop in plants with severe powdery mildew, exposing fruit to the sun.

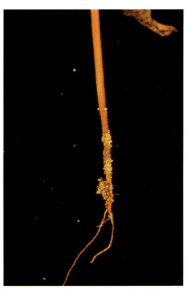

39. Lesion at the crown and white mycelial growth of *Sclerotium rolfsii* on the stem of a plant with southern blight.

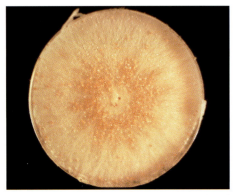

40. White mycelium and spherical sclerotia of *Sclerotium rolfsii* on a growth medium.

41. Early symptoms of Verticillium wilt caused by *Verticillium dahliae*.

42. Field symptoms of Verticillium wilt caused by *Verticillium dahliae*. (Courtesy D. L. Lindsey)

43. Vascular discoloration caused by *Verticillium dahliae*. (Courtesy E. Shannon)

44. Premature defoliation in a severe infection by *Verticillium dahliae.*

45. Zonate stem lesion caused by *Sclerotinia sclerotiorum.*

46. White mycelium of *Sclerotinia sclerotiorum* on an infected stem. (Courtesy G. W. Simone)

47. Sclerotia of *Sclerotinia sclerotiorum* in a stem.

48. Blotchy yellow and white mosaic induced on leaves by *Alfalfa mosaic virus.* (Courtesy E. Shannon)

49. Yellow mottle of Tabasco pepper leaves caused by the pepper strain of *Andean potato mottle virus.*

50. Yellow stipple and mottle of leaves of *Capsicum chinense* infected with the pepper strain of *Andean potato mottle virus.*

51. Young plant stunted by infection with *Beet curly top virus,* with yellow, thickened, crisp, rolled leaves.

52. Dark green veinbanding and leaf distortion caused by *Chilli veinal mottle virus* at a late stage of infection.

53. Distorted leaves of a mature plant infected with *Chino del tomate virus.*

54. Necrotic specks and ring spots on leaves and irregular ripening of fruit caused by *Cucumber mosaic virus.* (Courtesy T. A. Zitter)

55. Necrotic line pattern on leaves caused by *Cucumber mosaic virus.* (Courtesy T. A. Zitter)

56. Severely stunted plant infected with *Cucumber mosaic virus.* (Courtesy T. A. Zitter)

57. Bright golden foliar mosaic in infection by a Serrano golden mosaic virus–like isolate and the mosaic strain of Texas pepper virus. (Reprinted, by permission, from Brown and Bird, 1992)

58. Stunted leaves and crinkled leaf veins of a plant infected with *Pepper mild tigre virus* and the distortion strain of Texas pepper virus. (Courtesy D. Stenger)

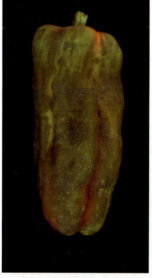

59. Mottled, malformed, off-colored fruit of a plant infected with *Pepper mild mottle virus.*

60. Systemic mottle caused by *Pepper mottle virus.* (Courtesy T. A. Zitter)

61. Vein mottling, mosaic, and leaf puckering caused by *Pepper veinal mottle virus.*

62. Systemic leaf mosaic caused by *Potato virus Y.*

63. Apical necrosis and associated symptoms of *Potato virus Y* infection.

64. Necrotic spots and deformation of fruit caused by *Potato virus Y.*

65. Leaf curl, interveinal chlorosis, and mosaic caused by *Sinaloa tomato leaf curl virus.*

66. Mosaic and leaf distortion caused by *Tobacco etch virus.* (Courtesy T. A. Zitter)

67. Mosaic and misshapen fruit of plants infected with *Tobacco etch virus.* (Courtesy T. A. Zitter)

68. Chlorotic mosaic and leaf distortion caused by *Tomato mosaic virus.*

69. Necrotic local lesions caused by *Tomato mosaic virus* on inoculated leaves of a variety resistant to the virus.

70. Stunting and chlorotic flecking of transplants infected with *Tomato spotted wilt virus.*

72. Necrotic ring spots on leaves caused by *Tomato spotted wilt virus.* (Courtesy R. Johnson)

73. Ring patterns and mosaic on fruit caused by *Tomato spotted wilt virus.*

71. Chlorotic and necrotic flecking on leaves of a plant infected with *Tomato spotted wilt virus.*

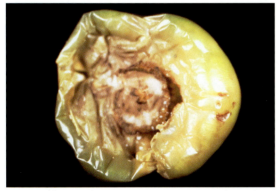

74. Fruit with bacterial soft rot invaded by secondary fungi. (Courtesy P. D. Roberts)

75. Sunken lesion covered with mold of the Alternaria rot pathogen.

76. Fruit damaged by *Botrytis cinerea,* with extensive fungal development.

77. Botrytis fruit rot of peppers in storage.

78. Dodder on field-grown pepper.

79. Stunting and chlorosis caused by *Meloidogyne incognita* in chile pepper.

80. Root galls caused by root-knot nematodes.

81. Root-knot nematode inside a root radicle, with an egg sac protruding at the surface.

82. Broad mite feeding damage to foliage.

83. Broad mite feeling damage to fruit.

84. Thrips feeding damage to fruit.

85. Thrips feeding damage to foliage.

86. Stink bug injury. (Courtesy T. A. Zitter)

87. Flat pod (right), with a normal fruit.

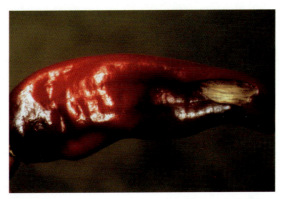

88. Blossom-end rot of chile pepper. (Courtesy E. Shannon)

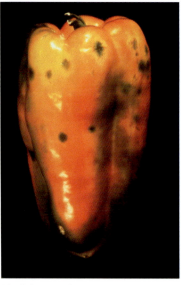

89. Blossom-end rot of bell peppers. (Courtesy T. A. Zitter)

90. Color spotting.

91. Flower and bud drop. (Courtesy C. Wien)

92. Pepper foliage damaged by hail.

93. Pimento pepper fruit damaged by hail. (Courtesy E. Shannon)

94. Seedling loss due to salt injury.

95. Poor stand establishment as a result of high salt concentration in the soil and other environmental factors.

COLOR PLATES

96. Salt injury in mature plants. (Courtesy E. Shannon)

97. Sunscald on foliage.

98. Sunscald on fruit. (Courtesy E. Shannon)

99. Sunscalded fruit with secondary fungal infection. (Courtesy E. Shannon)

100. Seedlings dying as a result of wind injury.

101. Foliage damaged by wind

102. Hypocotyl injury caused by wind.

103. Seedlings protected from wind by straw stubble.

104. Halosulfuron injury from direct spraying of foliage.

105. Imazethapyr injury from foliar exposure of bell pepper.

106. Pyrithiobac injury from direct spraying of chile pepper foliage.

107. Atrazine injury from soil exposure of bell pepper. (Courtesy T. A. Zitter)

108. Hexazinone injury from soil exposure of bell pepper. (Courtesy T. Lanini and C. Elmore)

109. Norflurazon injury from soil exposure of bell pepper.

110. Clomazone injury from soil exposure of bell pepper.

111. Oxyfluorfen injury from foliar exposure of bell pepper. (Courtesy T. Lanini and C. Elmore)

112. Carfentrazone injury from foliar exposure of bell pepper.

113. Paraquat injury from foliar exposure of chile pepper.

114. Bentazon injury from direct spraying of chile pepper foliage.

115. Glyphosate injury four days after direct spraying of chile pepper foliage with a 2% solution.

116. Glyphosate symptoms 30 days after foliar exposure of chile pepper.

117. Dicamba injury from foliar exposure of chile pepper.

118. Quinclorac injury from foliar exposure of bell pepper.

119. Necrosis of leaf margins and lower leaves due to potassium deficiency.

120. Stunting of plants due to potassium deficiency.

121. Interveinal chlorosis of middle and older leaves due to magnesium deficiency.

122. Light green leaves, a symptom of sulfur deficiency.

the effective resistance to CMV in mature plants, should reduce losses caused by the virus. These practices include the use of reflective mulches to deter aphid vectors; elimination of weeds in the vicinity of the crop; and timely planting, to avoid exposing young plants in the field when aphid populations are high, and particularly when aphids are in migration. Live mulches or border crops that are not susceptible to CMV may serve as depositories for the virus by viruliferous aphids entering the crop. Insecticides are not recommended, because of environmental concerns and their general lack of efficiency. In the nonpersistent manner of transmission of CMV by its aphid vectors, transmission occurs too rapidly for effective insecticide treatment. However, the combined effect of oil sprays and a rapid-acting insecticide (e.g., a pyrethroid) may reduce losses if applied in a timely manner.

Alternative approaches to managing CMV have been reported. Cross-protection, involving the inoculation of young plants with a mild strain of the virus or with the virus and an associated satellite RNA, has shown promise in pepper. Transgene-mediated protection against CMV has worked in several vegetable crops and may afford similar levels of protection in pepper. The application of root-colonizing bacteria that induce systemic resistance against CMV has been successful in tomato and may protect pepper in a similar manner.

Selected References

Agrios, G. N., Walker, M. E., and Ferro, D. N. 1985. Effect of cucumber mosaic virus inoculation at successive weekly intervals on growth and yield of pepper (Capsicum annuum) plants. Plant Dis. 69:52–55.

Gallitelli, D. 2000. The ecology of Cucumber mosaic virus and sustainable agriculture. Virus Res. 71:9–21.

Palukaitis, P., Roossinck, M. J., Dietzgen, R. G., and Francki, R. I. B. 1992. Cucumber mosaic virus. Adv. Virus Res. 41:281–348.

Tomlinson, J. A. 1987. Epidemiology and control of virus diseases of vegetables. Ann. Appl. Biol. 110:661–681.

(Prepared by J. F. Murphy)

Pepper golden mosaic virus

Since the late 1980s, epidemics caused by whitefly-transmitted geminiviruses have been prominent in pepper crops in Mexico, Central America, and the U.S. border states of Arizona and Texas. Serrano golden mosaic virus (SGMV) and *Pepper mild tigre virus* (PMTV), the latter now known to be a distinct begomovirus species, and previously unidentified begomoviruses associated with *tigre* symptoms in pepper (bright yellow chlorosis, mottling, and distortion of leaves; overall stunting; and poor fruit set and fruit distortion) have been reported in Mexico and the southern United States. When first described, SGMV and PMTV could be distinguished from other viruses known at the time by their distinct host ranges and symptoms in pepper. During the same period, other apparently unique begomoviruses were identified in pepper. The provisional name "Texas pepper virus" (TPV) was assigned to two strains from the Rio Grande Valley of Texas, the distortion strain (TPV-D) and the mosaic strain (TPV-Mo), which were distinguished by the symptoms they cause. Since then, isolates that are close relatives of TPV-D and TPV-Mo have been identified in diseased pepper in Texas, Mexico, and Central America, including Pepper virus isolates A, B, C, and D; SGMV; TPV-Tamaulipas (TPV-TAM); and TPV–Costa Rica (TPV-CR). The lack of data on the nucleotide sequences of these viruses, until recently, has prevented a definitive analysis of their relationships. This collection of isolates is now thought to be a complex of begomovirus variants, which has been designated *Pepper golden mosaic virus* (PepGMV).

Symptoms

The occurrence of symptoms suggests that members of the PepGMV complex are widely distributed, and disease incidence is high where they are present. Infection reduces yields and yield quality, but the extent of these losses has not been determined. Infected pepper plants exhibit a range of symptoms, probably depending on the composition of the complex and the pepper species or cultivar. A complex consisting of an SGMV-like isolate and TPV-Mo causes foliar mosaic ranging from dull yellow to strikingly bright gold (Plate 57). In contrast, a complex of a PMTV-like isolate and TPV-D causes yellow vein etching, leaf stunting, and crinkling or distortion of leaf veins, but less foliar discoloration and greater overall stunting of plants (Plate 58). However, sequence comparisons with the recently cloned DNA-A component of PMTV suggest that it is a separate species and, further, that PMTV symptoms may have been confounded by its occurrence together with the PepGMV complex. The original PMTV isolate was distinguished from the others by the yellow green splotches it caused on inoculated leaves and subsequent moderate distortion and stunting. On highly symptomatic plants, the fruit is often misshapen, discolored, and smaller than normal. Symptoms vary widely among pepper species and cultivars.

Causal Agent

PepGMV is thought to be a complex of closely related New World genotypes in the genus *Begomovirus*, family *Geminiviridae*, occurring individually or in mixtures. Members of the complex have a bipartite genome consisting of approximately 5.2 kb. The two genomic components, A and B, contain six or seven genes encoding proteins needed for the virus to complete an infection cycle. Studies in progress involving cloning and sequencing of full-length genomic components and the completion of Koch's postulates suggest that TPV-D, TPV-Mo, field isolates of TPV, and several field isolates of SGMV are variant strains of a single begomovirus species. Comparisons of viral sequences indicate that TPV-TAM and TPV-CR are also strains or variants of this species complex.

Members of the PepGMV complex are transmitted by the whitefly *Bemisia tabaci* in a persistent manner, with the whitefly being capable of transmitting the virus for at least 10 days after exposure to an infected plant, and possibly for the rest of its life. Natural hosts of the virus complex include pepper, tobacco, and tomato. *Datura stramonium* and *Nicotiana benthamiana* are experimental hosts. Bean, eggplant, *Malva parviflora*, and soybean are not susceptible to members of the complex.

Some isolates of the PepGMV complex are mechanically transmitted from pepper to pepper with great difficulty, while others have not been shown to be mechanically transmissible in pepper or in tomato. Members of the PepGMV complex are not seed-transmitted.

Disease Cycle and Epidemiology

There is little information on the disease cycle of the PepGMV complex under field conditions. The disease is prominent when the whitefly vector is present, and it seems to be more severe when tomatoes are grown near peppers, perhaps because the whitefly typically prefers to breed on tomato plants. Because the whitefly may transmit the virus for its lifetime, the virus complex poses a threat to solanaceous crops throughout the distribution area of the vector. There is no information on weeds as reservoirs of these begomovirus variants.

Control

To limit disease spread, it is essential to reduce or eliminate the whitefly vector from pepper and nearby vegetable fields where it breeds. Transplanting clean seedlings is the best method of control. Weed control near fields and removal of infected crops prior to planting the next crop are advised. There are no resistant pepper cultivars.

Selected References

Bonilla-Ramirez, G. M., Guevara-Gonzalez, R. G., Garzon-Tiznado, J. A., Ascenco-Ibanez, J. T., Torres-Pacheco, I., and Rivera-Bustamante, R. F. 1997. Analysis of the infectivity of monomeric clones of pepper huasteco virus. J. Gen. Virol. 78:947–951.

Brown, J. K. 2001. Molecular markers for the identification and global tracking of whitefly vector–begomovirus complexes. Virus Res. 71:233–260.

Brown, J. K., and Bird, J. 1992. Whitefly-transmitted geminiviruses and associated disorders in the Americas and the Caribbean Basin. Plant Dis. 76:220–225.

Brown, J. K., Idris, A. M., Torres-Jerez, I., Banks, G. K., and Wyatt, S. D. 2001. The core region of the coat protein gene is highly useful for establishing the provisional identification and classification of begomoviruses. Arch. Virol. 146:1–18.

Brown, J. K., and Nelson, M. R. 1989. Two whitefly-transmitted geminiviruses isolated from pepper affected with tigré disease. (Abstr.) Phytopathology 79:908.

Brown, J. K., and Poulos, B. T. 1990. Serrano golden mosaic virus: A newly identified whitefly-transmitted geminivirus of pepper and tomato in the United States and Mexico. Plant Dis. 74:720.

Brown, J. K., Campodonico, O. P., and Nelson, M. R. 1989. A whitefly-transmitted geminivirus from peppers with tigré disease. Plant Dis. 73:610.

Lotrakul, P., Valverde, R. A., De La Torre, R., Sim, J., and Gomez, A. 2000. Occurrence of a strain of *Texas pepper virus* in Tabasco and Habanero pepper in Costa Rica. Plant Dis. 84:168–172.

Stenger, D. C., Duffus, J. E., and Villalon, B. 1990. Biological and genomic properties of a geminivirus isolated from pepper. Phytopathology 80:704–709.

Torres-Pacheco, I., Garzón-Tiznado, J. A., Brown, J. K., Becerra-Flora, A., and Rivera-Bustamante, R. F. 1996. Detection and distribution of geminiviruses in Mexico and the southern United States. Phytopathology 86:1186–1192.

Torres-Pacheco, I., Garzón-Tiznado, J. A., Herrera-Estrella, L., and Rivera-Bustamante, R. F. 1993. Complete nucleotide sequence of pepper huasteco virus: Analysis and comparison with bipartite geminiviruses. J. Gen. Virol. 74:2225–2231.

(Prepared by J. K. Brown)

Pepper huasteco yellow vein virus

A whitefly-transmitted geminivirus species detected in pepper plants in Mexico was recently described and designated *Pepper huasteco yellow vein virus* (PHYVV) (syn. Pepper huasteco virus). Comparison of the complete nucleotide sequence of the A and B components of the PHYVV genome with the sequences of other well-studied begomoviruses provides convincing evidence, reinforced by the distinct symptomatology of the virus, that PHYVV is a distinct begomovirus species. A survey of pepper fields in Mexico and U.S. border states indicated that PHYVV and the *Pepper golden mosaic virus* (PepGMV) complex are among the most widespread and economically important begomoviruses infecting pepper in the region. Yields of pepper plants infected with PHYVV are believed to be reduced, but no definitive data on yield loss are available.

Symptoms

PHYVV can be distinguished from other begomoviruses by its nucleotide sequence and unique symptoms in pepper and tomato. Symptoms in pepper are particularly striking in that veins become bright yellow upon inoculation, and infected leaves exhibit a diffuse mosaic. Fruit set is reduced and plants are stunted when infected with this virus.

Causal Agent

PHYVV is a member of the genus *Begomovirus*, in the family *Geminiviridae*, with a bipartite genome of approximately 5.2 kb, encoding six or seven proteins. Comparisons with full-length nucleotide sequences of other begomoviruses indicate that PHYVV is only distantly related to other New World pepper-infecting begomoviruses, including *Chino del tomate virus, Sinaloa tomato leaf curl virus,* and the PepGMV complex (strains and field isolates of Texas pepper virus and Serrano golden mosaic virus). PHYVV is in a lineage separate from all of these viruses (*unpublished data,* J. K. Brown and D. C. Stenger).

PHYVV is a typical bipartite New World begomovirus, with the exception that its replicase-associated protein (Rep) resembles that of several Old World begomoviruses more than that of New World species. It appears to be the first reported member of a distinctive begomovirus clade from Mexico and the Caribbean region. PHYVV appears to have a narrow host range, but little information on the extent of natural or experimental hosts is available. It is transmitted in a persistent manner by the whitefly *Bemisia tabaci* and can be transmitted for the life of the vector. There is no evidence that PHYVV is seed-transmitted.

Disease Cycle and Epidemiology

There is little information concerning the epidemiology of PHYVV in pepper, although it is known to occur even when whitefly populations are low. The virus is apparently transmitted by local whitefly populations and by *B. tabaci* biotype B, which is present along the east coast of Mexico, where PHYVV was originally found.

Control

It is essential to reduce or eliminate the whitefly population in pepper and adjacent crops in which the vector breeds. The use of virus-free transplants is helpful, but once plants become infected, some damage probably occurs. Pepper is the least damaged of the hosts of the virus, however, so that rouging infected plants can reduce inoculum in pepper fields, when tomato is not planted nearby. Weed control near pepper fields and removal of an infected crop prior to planting a subsequent crop are advised. There is one report of possible resistance in pepper, by which viral movement within the plant is restricted.

Selected References

Brown, J. K. 2001. Molecular markers for the identification and global tracking of whitefly vector–begomovirus complexes. Virus Res. 71:233–260.

Brown, J. K., Idris, A. M., Torres-Jerez, I., Banks, G. K., and Wyatt, S. D. 2001. The core region of the coat protein gene is highly useful for establishing the provisional identification and classification of begomoviruses. Arch. Virol. 146:1–18.

Garzón-Tiznado, J. A., Torres-Pacheco, I., Ascencio-Ibañez, J. T., Herrera-Estrella, L., and Rivera-Bustamante, R. F. 1993. Inoculation of peppers with infectious clones of a new geminivirus by a biolistic procedure. Phytopathology 83:514–521.

Godinez-Hernandez, Y., Anaya-Lopez, J. L., Diaz-Plaza, R., Gonzalez-Chavira, M., and Torres-Pacheco, I. 2001. Characterization of resistance to *Pepper huasteco* geminivirus in chili peppers from Yucatan, Mexico. HortScience 36:139–142.

Idris, A. M., Lee, S. H., and Brown, J. K. 1999. First report of chino del tomate and pepper huasteco geminiviruses in greenhouse-grown tomato in Sonora, Mexico. Plant Dis. 83:396.

(Prepared by J. K. Brown)

Pepper mild mottle virus

Pepper mild mottle virus (PMMV) occurs worldwide in chili and bell pepper plants. It drastically reduces marketable fruit yield and is consequently of great economic importance. Severe outbreaks have been reported in Spain and Sicily in peppers grown under structures such as plastic tunnels and in greenhouses. Mixed infections of PMMV with other tobamoviruses, such as *Tobacco mosaic virus* (TMV) and *Tomato mosaic virus* (ToMV), are common in farmers' fields.

Symptoms

PMMV usually causes only mild symptoms, such as mottle and green or yellow mosaic, on the leaves of pepper plants. Symptoms on fruit, however, tend to be pronounced. Fruit may be small, mottled, and malformed, with sunken or raised areas that can be off-colored, necrotic, or both (Plate 59). The symptoms vary greatly in intensity according to the viral strain, pepper cultivar, and time of infection (growth stage and season).

Causal Agent

PMMV is a member of the genus *Tobamovirus*. The virions are rigid, rod-shaped particles, approximately 312×18 nm, which are extremely stable and can persist in the environment for long periods of time. Its particle stability and ability to accumulate to high levels in infected tissues make PMMV easily sap-transmissible and readily transferred from plant to plant by workers' hands and tools during cultivation. The virus typically induces angled-layer aggregates in the cytoplasm of infected pepper plants.

The host range of PMMV is primarily in the Solanaceae. Chili and bell peppers are its only reported natural hosts. Experimentally, the virus can infect various *Nicotiana* spp. locally or systemically. It induces local lesions on inoculated leaves of *Chenopodium quinoa* and *Datura stramonium*. Petunia is a symptomless host. PMMV does not infect *Eryngium planum*, an experimental host of *Tobacco mild green mosaic virus* (TMGMV), a tobamovirus that has occasionally been found in pepper plants in the United States and Italy. Tomato is not a host of PMMV or TMGMV, which distinguishes these viruses from ToMV and TMV. Both of the latter two infect peppers worldwide.

Entire genome analysis has shown that PMMV is located in the same cluster as ToMV, TMV, and TMGMV and is most closely related to ToMV and TMV. Serological differentiation of these four tobamoviruses is possible by double-antibody sandwich enzyme-linked immunosorbent assay (ELISA).

Four strains of PMMV have been reported (P_0, P_1, $P_{1.2}$, and $P_{1.2.3}$), which can be distinguished by the hypersensitive reactions they induce in *Capsicum* genotypes bearing the L^1, L^2, L^3, and L^4 resistance genes in the homozygous condition. The PMMV strains cannot be differentiated serologically but can be distinguished by reverse transcriptase polymerase chain reaction (RT-PCR) followed by restriction enzyme digestion.

Disease Cycle and Epidemiology

PMMV has no known biological vector. Infected plant debris in the soil and contaminated seed are common sources of primary infection in the field. The virus can survive several months in the soil in infected leaves, stems, and roots. Seed transmission in pepper plants varies greatly and can be as high as 30%. The virus occurs on the outer and inner seed coat and on the inseparable embryo and endosperm. Handling and touching plants during standard cultivation can spread PMMV within a planting, especially in protected cultivation.

Control

Planting clean seed is one of the most important measures to avoid initial infection by PMMV. The virus can usually be eliminated from the external surface of seed by treatment with a 10% solution of trisodium phosphate (TSP, Na_3PO_4) for 20 min, at a ratio of 250 ml of TSP per 100 g of seed. A 2.5-hr treatment with 10% TSP may inactivate PMMV in the internal portions of the seed as well as on the external surface. The treatment is most effective if the TSP is changed once after 30 min and the seeds are stirred in the solution during treatment. A thorough rinse in tap water after the treatment is essential. Neither treatment should affect germination. Dry heat treatment appears to be ineffective in eliminating internal seedborne PMMV, and it will adversely affect germination. Seedlings for transplanting should be raised from virus-free seed and sown in sterilized soil, preferably in new seedling trays. Since PMMV can survive in plant debris for several months or even longer, seedlings should not be transplanted into fields where peppers have recently been grown.

While preventive seed treatment is an important approach to PMMV management, genetic resistance offers the most effective control. PMMV-resistant cultivars carrying the L^2 and L^3 genes and the recently identified L^4 gene are commercially available. The L^3 and L^4 resistance alleles originate from *C. chinense* and *C. chacoense*, respectively.

Selected References

Alonso, E., Garcia-Luque, I., Avila Rincon, M. J., Wicke, B., Serra, M. T., and Diaz-Ruiz, J. R. 1989. A tobamovirus causing heavy losses in protected pepper crops in Spain. J. Phytopathol. 125:67–76.

Alonso, E., Garcia-Luque, I., de la Cruz, A., Wicke, B., Avila-Rincon, M. J., Serra, M. T., Castresana, C., and Diaz-Ruiz, J. R. 1991. Nucleotide sequence of the genomic RNA of pepper mild mottle virus, a resistance-breaking tobamovirus in pepper. J. Gen. Virol. 72:2875–2884.

Green, S. K., and Kim, J. S. 1991. Characteristics and control of viruses infecting peppers: A literature review. Asian Veg. Res. Dev. Cent. Tech. Bull. 18.

Rast, A. Th. B. 1988. Pepper tobamoviruses and pathotypes used in resistance breeding. Capsicum Newsl. 7:20–23.

Rast, A. Th. B., and Stijger, C. C. M. M. 1987. Disinfection of pepper seed infected with different strains of capsicum mosaic virus by trisodium phosphate and dry heat treatment. Plant Pathol. 36:583–588.

Tenllado, F., Garcia-Luque, I., Serra, M. T., and Diaz-Ruiz, J. R. 1994. Rapid detection and differentiation of tobamoviruses infecting L-resistant genotypes of pepper by RT-PCR and restriction analysis. J. Virol. Methods 47:165–174.

Wetter, C. 1984. Serological identification of four tobamoviruses infecting pepper. Plant Dis. 68:597–599.

Wetter, C., and Conti, M. 1988. Pepper mild mottle virus. Descriptions of Plant Viruses, no. 330. Commonwealth Mycological Institute and Association of Applied Biologists, Kew, England.

(Prepared by S. K. Green)

Pepper mottle virus

Pepper mottle virus (PepMoV) appears to be limited to the Western Hemisphere, primarily the southern United States and California, Mexico, and Central America. It was initially described concurrently in Florida, where it induced severe losses in bell pepper production, and in Arizona, where it affected chili peppers. In Florida and Arizona, the causal agent was shown to be closely related to *Potato virus Y* (PVY), but was later identified as a distinct virus.

Symptoms

In a controlled environment, such as in a greenhouse, faint and diffuse chlorotic lesions may develop on mechanically inoculated leaves of some pepper varieties. Younger, uninoculated leaves initially develop moderate to pronounced vein-clearing followed in these leaves and primarily in younger leaves by a distinct mottle. Mottled young leaves tend to be smaller than comparable leaves of healthy plants and are rather brittle. In *Capsicum frutescens* 'Tabasco,' PepMoV induces necrotic local lesions on inoculated leaves. Some isolates of the virus also cause subsequent systemic necrosis and apical shoot death. Regrowth of infected foliage below a dead shoot may occur.

The chlorotic lesions on inoculated leaves and systemic vein-clearing of greenhouse-grown pepper plants are typically not apparent on plants grown and inoculated in the field. A characteristic systemic mottle, however, is clearly identifiable under field conditions (Plate 60). Plants infected at an early stage of development may be severely stunted and have small, misshapen young leaves. If plants are closer to maturity at the time of infection, a systemic mottle may develop, with little, if any, stunting and leaf deformation.

PepMoV causes severe systemic symptoms in *Nicotiana tabacum* 'Xanthi' and 'Burley'-related varieties and in *N. benthamiana*. Some isolates have been reported to induce chlorotic local lesions in *Chenopodium amaranticolor,* but inoculum from pepper extract may not induce this response. The host range of the virus is essentially limited to solanaceous species.

Causal Agent

PepMoV is a member of the genus *Potyvirus,* in the family *Potyviridae.* Particles of this virus are filamentous, flexuous rods, approximately 737 nm long. The PepMoV genome consists of a single molecule of single-stranded, messenger-sense RNA, 9,640 nucleotides in length (excluding the poly A tail), which encodes a single polyprotein of 3,068 amino acids. As in all members of the genus *Potyvirus,* individual proteins are cleaved from the polyprotein by autocatalytic cleavage from proteinases that reside within the polyprotein.

Amino acid sequence homology of PepMoV and other potyviruses reveals that PVY has the highest degree of similarity to PepMoV. Although these two viruses are the most closely related, they can be distinguished in several ways. Biologically, they can be distinguished by differences in symptomatology and host range, including differences in the responses of key species of pepper. Serologically, the two viruses are distinguished by properties of the coat protein and cylindrical inclusion protein. PepMoV induces the formation of pinwheel-shaped cylindrical inclusions (CI) in the cytoplasm of infected cells. CI induced by PepMoV, in contrast to those induced by PVY, appear prominently and more abundantly at the cell wall during early stages of infection, and they tend to be long and thin, whereas PVY CI tend to be shorter. Furthermore, during later stages of infection when CI accumulate in the cytoplasm, PepMoV CI appear to be much larger than those formed in PVY infections.

Three PepMoV strains (Arizona, California, and Florida) have been documented. The Arizona and Florida strains have been well characterized biologically and serologically, and the complete viral RNA nucleotide sequence is available for the California and Florida strains. Other variants have been described.

Disease Cycle and Epidemiology

PepMoV is transmitted in a nonpersistent manner by nymphs and adults of *Myzus persicae.* Other aphid species also serve as vectors, but *M. persicae* is considered the most efficient. It is likely that PepMoV overwinters in weed species. *Datura meteloides* has been considered the primary overwintering host in Arizona, and *Solanum nigrum* in southern Florida, although no extensive evaluation of other potential weed hosts has been reported. Seed transmission of PepMoV did not occur in a limited number of *Capsicum* cultivars tested, and no information on seed transmission in weed species is available.

Mixed infections of PepMoV, PVY, and TEV frequently occur in the field. These viruses can be distinguished from one another by inoculation of indicator hosts and by serological procedures.

Control

Genotypes resistant to PepMoV have been described, but only Delray Bell has been used commercially. A more recent release, *C. annuum* 'Dempsey,' has been reported to be resistant to both the California strain of PepMoV and *Tobacco etch virus.*

Because of the nonpersistent manner of transmission of the virus by aphids, it is difficult to manage PepMoV by applying insecticides to control the vectors. However, application of mineral oil in combination with synthetic pyrethroids has been shown to reduce PVY incidence in pepper and may offer similar protection against PepMoV. Reflective mulches to deter insects may delay infection, thereby reducing disease severity and yield losses.

As in the management of most virus diseases, isolation of fields from potential sources of inoculum and weed control in the vicinity of pepper fields may limit the introduction of the virus and reduce subsequent yield losses.

Selected References

Hiebert, E., and Purcifull, D. E. 1992. A comparison of pepper mottle virus with potato virus Y and evidence for their distinction. Arch. Virol. (Suppl. 5):321–326.

Nelson, M. R., and Wheeler, R. E. 1972. A new virus disease of pepper in Arizona. Plant Dis. Rep. 56:731–735.

Nelson, M. R., Wheeler, R. E., and Zitter, T. A. 1982. Pepper mottle virus. Descriptions of Plant Viruses, no. 253. Commonwealth Mycological Institute and Association of Applied Biologists, Kew, England.

Vance, V. B., Moore, D., Turpen, T. H., Bracker, A., and Hollowell, V. C. 1992. The complete nucleotide sequence of pepper mottle virus genomic RNA: Comparison of the encoded polyprotein with those of other sequenced potyviruses. Virology 191:19–30.

Warren, C. E., and Murphy, J. F. 2003. The complete nucleotide sequence of *Pepper mottle virus*–Florida RNA. Arch. Virol. 148:189–197.

Zitter, T. A. 1972. Naturally occurring pepper virus strains in south Florida. Plant Dis. Rep. 56:586–590.

(Prepared by J. F. Murphy and T. A. Zitter)

Pepper veinal mottle virus

Pepper veinal mottle virus (PVMV) was first reported in West Africa in 1971 and has since been isolated in many other parts of Africa, the Middle East, India, and southeast Asia. Wherever the virus occurs, it causes great losses in yield and quality of *Capsicum* species.

Symptoms

PVMV is primarily a pathogen of pepper, but it also infects other solanaceous crops. In *Capsicum annuum* and *C. frutescens,* the symptoms are vein mottling, mosaic, and leaf deformations with puckering (Plate 61). Veinal chlorosis may be vivid at times. Plants are usually stunted, and the stunting contributes to the lack of fruit set and the production of distorted fruit.

Isolates of PVMV differ in host range and pathogenicity. Some, but not all, infect eggplant, tomato, and some species of tobacco. In tomato, PVMV isolates may induce necrotic lesions on inoculated leaves, followed by systemic leaf mottling; necrosis of leaf, stem, and petioles (streak); and leaf abscission and stunting. In eggplant, the PVMV type isolate causes systemic chlorotic spotting and a mild mottle. However, some isolates produce faint chlorotic local lesions with no systemic infection, and inoculation with other PVMV isolates results in no apparent infection. In *Nicotiana benthamiana,* PVMV causes veinclearing, mottling, and eventually plant death. In *N. clevelandii* and *N. glutinosa,* the virus causes conspicuous veinclearing and systemic mottle, but some isolates fail to infect *N. tabacum* 'White Burley' and other Burley types. In *N. tabacum* 'Xanthi,' most isolates of PVMV produce whitish spots or chlorotic areas on inoculated leaves, often after 25–30 days, but the virus does not move systemically. The movement of PVMV in Xanthi plants is not as restricted, however, when the plants are coinfected with PVY.

Causal Agent

PVMV is member of the genus *Potyvirus,* in the family *Potyviridae.* It is a single-stranded RNA virus with filamentous particles, 770 × 12 nm. The virus induces type 4 cylindrical inclusions, consisting of pinwheels, scrolls, and short, curved, laminated aggregates, and is assigned to Subdivision IV of the *Potyviridae.* PVMV also induces irregularly shaped cytoplasmic inclusions of aggregated tubules and electron-opaque material, as do *Pepper mottle virus* (PepMoV) and *Potato virus Y* (PVY). The PVMV-induced tubules are straight, resembling those caused by PVY, but are distinct from the convoluted tubules induced by PepMoV.

PVMV is serologically distinct from PVY and other pepper-infecting potyviruses. PVMV isolates appear to be very homogeneous serologically, and no significant differences in epitopes were found when isolates from seven West African countries were tested. However, another study conducted in West Africa revealed two pathogroups. PVMV strains have been identified on the basis of host range and symptoms induced, mostly in solanaceous plants.

Disease Cycle and Epidemiology

PVMV is transmitted in a nonpersistent manner by at least eight aphid species. *Myzus persicae, Aphis gossypii, A. craccivora,* and *A. spiraecola* have been described as efficient vectors in nature. Weed hosts probably play a key role in maintaining the virus over the summer, while also serving as suitable hosts for the vectors. *Physalis angulata, P. micrantha,* and *Solanum nigrum* are reported to be hosts of PVMV in Africa. The virus has not been reported to be seedborne in pepper or any other crop or weed host. Mixed infections by PVMV with other viruses, including PVY, have been reported in Africa and elsewhere.

Control

Because of the widespread importance of PVMV, considerable efforts have focused on the development of pepper varieties resistant to the virus. Early work resulted in materials that were tolerant or only partially resistant. Recently, a doubled haploid (DH) line, DH801, with complete resistance to PVMV was developed, originating from an F_1 hybrid of the *C. annuum* cultivars Perennial and Florida VR2. A recessive gene from Perennial was complementary with *pvr2* (from Florida VR2, conferring resistance to PVY 0 and 1) and was tentatively designated *pvr6*. With different pathotypes of PVMV having been reported, additional tests under field conditions will be required to determine the durability of this resistance.

With the nonpersistent manner of transmission of PVMV and its extensive host range among solanaceous crops, attempts to manage the disease by the use of insecticides and even mineral oil sprays are likely to be ineffectual. Isolation of fields and planting tall barrier crops may delay the early introduction of virus and reduce the subsequent yield loss.

Selected References

Brunt, A. A., and Kenten, R. H. 1971. Pepper veinal mottle virus, a new member of the potato virus Y group from peppers (*Capsicum annuum* L. and *C. frutescens* L.) in Ghana. Ann. Appl. Biol. 69: 235–243.

Brunt, A. A., and Kenten, R. H. 1972. Pepper veinal mottle virus. Descriptions of Plant Viruses, no. 104. Commonwealth Mycological Institute and Association of Applied Biologists, Kew, England.

Brunt, A. A., Kenten, R. H., and Phillips, S. 1978. Symptomatologically distinct strains of pepper veinal mottle virus from four West African solanaceous crops. Ann. Appl. Biol. 88:115–119.

Caranta, C., Palloix, A., Gebre-Selassie, K., Lefebvre, V., Moury, B., and Daubeze, A. M. 1996. A complementation of two genes originating from susceptible *Capsicum annuum* lines confers a new and complete resistance to pepper veinal mottle virus. Phytopathology 86:739–743.

Gorsane, F., Fakhfakh, H., Tourneur, C., Makni, M., and Marrakchi, M. 1999. Some biological and molecular properties of pepper veinal mottle virus isolates occurring in Tunisia. Plant Mol. Biol. Rep. 17: 149–158.

Hiskias, Y., Lesemann, D. E., and Vetten, H. J. 1999. Occurrence, distribution and relative importance of viruses infecting hot pepper and tomato in the major growing areas of Ethiopia. J. Phytopathol. 147:5–11.

Huguenot, C., Furneaux, M. T., Clare, J., and Hamilton, R. I. 1996. Serodiagnosis of pepper veinal mottle virus in West Africa using specific monoclonal antibodies in DAS-ELISA. J. Phytopathol. 144:29–32.

Igwegbe, E. C. K., and Waterworth, H. E. 1982. Properties and serology of a strain of pepper veinal mottle virus isolated from eggplant (*Solanum melongena* L.) in Nigeria. Phytopathol. Z. 103:9–12.

Konate, G., and Traore, O. 1999. Characterization and distribution of pepper veinal mottle virus in West Africa. Cah. Agric. 8:132–134.

Ladipo, J. L., and Roberts, I. M. 1977. Pepper veinal mottle virus associated with a streak disease of tomato in Nigeria. Ann. Appl. Biol. 87:133–138.

Marchoux, G., Delecolle, B., and Gebre-Selassie, K. 1993. Systemic infection of tobacco by pepper veinal mottle potyvirus (PVMV) depends on the presence of potato virus Y (PVY). J. Phytopathol. 137:283–292.

(Prepared by T. A. Zitter)

Potato virus Y

Potato virus Y (PVY) occurs worldwide in pepper. It was first identified in the United States in the 1940s and has since been reported many times in almost every pepper cultivation region in the world. Diseases caused by the virus are more common in open-field cultivation in warm climates. PVY is an important pathogen of other major solanaceous crops, such as potato, tobacco, tomato, and (to a lesser extent) eggplant and petunia. Its host range is mainly confined to plants in the Solanaceae, but it can also infect members of the Amaranthaceae, Asteraceae, Chenopodiaceae, and Fabaceae.

Symptoms

The most common symptoms of PVY infection in pepper are systemic veinclearing progressing into a mosaic or mottle and, generally, dark green veinbanding of the leaves (Plate 62). Vein and petiole necroses often develop. In some cases, these symptoms are followed by stem necrosis and defoliation, necrosis of the apical bud, and even plant death (Plate 63). Necrotic spots and mosaic patterns may develop on the fruit of some pepper cultivars, and the fruit may be deformed (Plate 64); however, fruit symptoms do not always occur in PVY-infected plants. Other symptoms reported in association with PVY infection of pepper include stunting, leaf distortion, a reduction in fruit size, and abortion of flowers.

The extent of PVY-induced losses in pepper crops varies, depending on factors such as the age of the plants at the time of infection, virus isolate, pepper cultivar, and environmental conditions. Infection at an early stage of plant development can result in yield losses of as much as 100%.

Pepper plants infected with PVY may also be infected with one or more distinctly different viruses, which may cause more severe symptoms and greater yield losses than those due to PVY alone. In many cases, symptoms induced by PVY are masked by the more severe symptoms caused by other viruses. Viruses that are commonly identified in coinfections with PVY include *Alfalfa mosaic virus, Broad bean wilt virus, Cucumber mosaic virus,* and *Tobacco etch virus.*

Causal Agent

PVY is the type member of the genus *Potyvirus,* in the family *Potyviridae.* This genus contains the largest number of distinct viruses. The PVY virion is a flexuous, rod-shaped particle, 740 × 11 nm, made up of about 2,000 copies of a single coat protein of approximately 35 kDa, helically arrayed. Each virion contains a single RNA molecule of approximately 9,700 nucleotides, which is covalently attached to a protein (VPg) at its 5′ terminus and polyadenylated at its 3′ terminus. PVY can be mechanically transmitted experimentally, and the RNA alone is infectious. Viral proteins are produced in infected plants by virus-encoded protease cleavage of a single polyprotein translated from the viral genomic RNA. Cytoplasmic pinwheel-like inclusions, which are characteristic structures associated with potyvirus infections, accumulate in infected cells.

Antibodies against PVY virions are easily obtained. The serological relationships among different strains and isolates of the virus may vary, depending on the particular antibodies used in the analysis. However, pepper-infecting isolates can be distinguished from the more typical potato isolates by means of monoclonal antibodies. Serological and molecular variability in the pepper-infecting PVY strains is more restricted than in other potato-infecting strains, but the variability is enough to allow the identification of several virus pathotypes able to overcome certain resistance genes. Antibodies directed against PVY proteins other than the coat protein, such as the cylindrical inclusion (CI) protein or the helper component proteinase (HC-Pro) protein, can also be used to characterize the relationships of PVY isolates and strains.

Disease Cycle and Epidemiology

The only known means by which PVY is spread in the field is nonpersistent transmission by aphids. No transmission by seed, pollen, or contact has been reported. Thus, two main factors affect the disease cycle and epidemiology: the presence of aphids and the presence of a reservoir of PVY. *Myzus persicae,* a pepper colonizer, is the most efficient vector, although it probably plays a major role only in the secondary dispersion of PVY when it arrives late in the cultivation season. Noncolonizing aphids visiting young pepper plants shortly after transplanting appear to play a major role in primary infection. Aphids in the genus *Aphis* are probably the most important vectors in this regard. In the nonpersistent mode of transmission, viruliferous noncolonizing aphids can inoculate plants with PVY in just seconds or minutes during feeding probes.

Virus reservoirs in weeds or cultivated crops, such as tomato and tobacco grown near pepper, may serve as sources of primary infection. The role of potato as an overwinter reservoir for subsequent infection of pepper is questionable, since most PVY isolates infecting potato normally do not infect pepper (likewise, most isolates from pepper do not infect potato). Weeds serving as PVY reservoirs include *Solanum nigrum* and other *Solanum* spp., *Portulaca oleracea,* and *Physalis* spp.

Control

Control measures for PVY in pepper follow two general approaches: prevention of infection through appropriate cultural practices and the use of resistant cultivars. Once infected, a susceptible plant cannot be cured, so roguing is advisable to diminish secondary spread to uninfected plants. Chemical control of aphids does not prevent primary infection, because noncolonizing aphids feed briefly and then take flight, transmitting the virus in the process. For control of secondary dispersion within a field, it is advisable to keep colonizing aphid populations low, although chemical treatment has little effect, because of the nonpersistent mode of transmission of PVY. Weed control can effectively lower virus reservoirs. Barrier crops have been tested, with promising results.

The most effective control measure is the use of resistant cultivars. Several genes for resistance to PVY have been introgressed into different pepper cultivars, and others have been identified in wild pepper germ plasm for eventual use in breeding programs. Substantial alleviation of the disease has been achieved in many pepper-producing areas of the world by the use of commercial cultivars carrying resistance genes, although new, resistance-breaking virus pathotypes have subsequently appeared. Current breeding efforts aim to use new, broad sources of resistance more durable than present ones.

Selected References

Fereres, A. 2000. Barrier crops as a cultural control measure of nonpersistently transmitted aphid-borne viruses. Virus Res. 71:221–231.

Fereres, A., Pérez, P., Gemeno, C., and Ponz, F. 1993. Transmission of Spanish pepper and potato-PVY isolates by aphid vectors: Epidemiological implications. Env. Entomol. 22:1260–1265.

Gebré Selassie, K., Marchoux, G., Delecolle, B., and Pochard, E. 1985. Variabilité naturelle des souches du virus Y de la pomme de terre dans les cultures de piment du sud-est de la France: Caractérisation et classification en pathotypes. Agronomie 5:621–630.

Green, S., and Kim, J. S. 1991. Characteristics and control of viruses infecting peppers: A literature review. Asian Veg. Res. Dev. Cent. Tech. Bull. 18.

Kyle, M., and Palloix, A. 1997. Proposed revision of nomenclature for potyvirus resistance in *Capsicum.* Euphytica 97:183–188.

Llave, C., Martínez, B., Díaz-Ruiz, J. R., and López-Abella, D. 1999. Serological analysis and coat protein sequence determination of potato virus Y (PVY) pepper pathotypes and differentiation from other PVY strains. Eur. J. Plant Pathol. 105:847–857.

Luis-Arteaga, M., Arnedo-Andrés, M., and Gil-Ortega, R. 1997. New potato virus Y pathotype in pepper. Capsicum and Eggplant Newsl. 16:85–86.

Pasko, P., Gil-Ortega, R., and Luis-Arteaga, M. 1996. Resistance to potato virus Y in peppers. Capsicum and Eggplant Newsl. 15:11–27.

Romero, A., Blanco-Urgoiti, B., Soto, M. J., Fereres, A., and Ponz, F. 2001. Characterization of typical pepper-isolates of PVY reveals multiple pathotypes within a single genetic strain. Virus Res. 79:71–80.

Shukla, D. D., Ward, C. W., and Brunt, A. A. 1994. The Potyviridae. CAB International, Wallingford, U.K.

Soto, M. J., Luis-Arteaga, M., Fereres, A., and Ponz, F. 1994. Limited degree of serological variability in pepper strains of potato virus Y as revealed by analysis with monoclonal antibodies. Ann. Appl. Biol. 124:37–43.

(Prepared by M. Luis-Arteaga and F. Ponz)

Sinaloa tomato leaf curl virus

Sinaloa tomato leaf curl virus (STLCV) causes a widespread, debilitating disease in pepper and tomato, which was first observed in tomato in Sinaloa, Mexico, in 1989. The causal agent was tentatively described as a whitefly-transmitted geminivirus, on the basis of whitefly transmissibility and DNA-DNA hybridization experiments. The Mexican isolate, STLCV-MX, can be distinguished from other pepper-infecting begomoviruses by its moderately narrow host range, which includes only cultivated and weed species in the Solanaceae and Malvaceae, and by the symptoms it causes in key hosts. STLCV causes symptomatic infection of pepper, tobacco, and tomato, all of which have known centers of diversity in the New World, whereas it causes a symptomless infection in eggplant, a species that originated in the Old World. Further, the virus can be transmitted in sap to *Nicotiana benthamiana,* although with difficulty, but not between pepper and tomato. Collectively, these characteristics distinguish STLCV from other begomoviruses of solanaceous hosts in the Western Hemisphere.

Symptoms

STLCV causes moderate to severe leaf curl, interveinal chlorosis, mild yellow to green yellow mosaic on systemically infected leaves (Plate 65), shortened internodes, and stunting. In some pepper varieties, necrosis develops on the blossom end of the fruit, and plants infected early in the season produce small or stunted fruit. STLCV-infected tomato plants develop foliar curling and chlorosis, a unique purpling of the abaxial side of leaves, and shortened internodes.

The disease caused by STLCV in pepper can easily be confused with diseases caused by other begomoviruses if evaluated solely on the basis of symptoms and transmission phenotypes. For example, the distortion strain of Texas pepper virus causes distortion and stunting in pepper and mild mosaic and stunting in tomato. Pepper huasteco virus infection causes severe leaf curling and mosaic in both pepper and tomato. Both viruses are more readily sap-transmitted than STLCV. *Pepper mild tigre virus* differs, insofar as it causes mild leaf curling in tomato and a bright yellow splotchy mottle in pepper and is not mechanically transmissible. Comparison of the disease phenotype of *Taino tomato mottle virus,* from Cuba, with that of STLCV also indicates that they are distinct viruses. *Tomato yellow mosaic virus,* from Venezuela, and *Tomato mottle virus,* from Florida, are both readily mechanically transmitted and cause various degrees of yellow mosaic and leaf curling in tomato but have not been reported to infect pepper.

Causal Agent

STLCV is a member of the genus *Begomovirus,* in the family *Geminiviridae.* STLCV-MX is a bipartite, New World begomovirus from Mexico, with a genome of approximately 5.2 kb, encoding six or seven viral proteins, like other begomoviruses. At least one isolate, STLCV-CR, has been identified from tomato in Costa Rica, and it shares more than 98% nucleotide sequence identity with STLCV-MX, tentatively indicating it is a strain or simply another isolate of the virus. Only predictions about the relationship of STLCV-MX to other begomoviruses are available, because the entire genome has not yet been sequenced, nor have infectious clones been obtained. Consequently, evidence that the virus is the causal agent of the disease is indirect. However, it is known that the A and B component common region sequences involved in replication and transcription have conserved regions of about 174 nucleotides with 97.7% sequence identity. This indicates that clones derived by polymerase chain reaction are cognate A and B viral components of STLCV. The 100% identity in replication-essential sequences and those involved in transcriptional regulation of the A and B common region permit the prediction that the viral components would be readily compatible if obtained as full-length infectious clones. Furthermore, analogous common region sequences are not reported for any other begomovirus described to date. Pairwise sequence comparisons of the CP gene of STLCV and that of 36 other geminiviruses indicated nucleotide and amino acid sequence identities ranging from 18 to 83.6% and from 11.1 to 90.1%, respectively. The CP nucleotide sequence of STLCV is most like that of *Chino del tomate virus* (CdTV), which is from the same region of Mexico, with 84% sequence identity. Although CdTV also has an origin on the west coast of Mexico, the CP is more like that of begomoviruses in the Abutilon mosaic group (or clade), which contains,

among others, *Abutilon mosaic virus* in the West Indies and *Tomato mottle virus* in Florida. Consequently, STLCV does not appear to be a strain of any of its closest relatives.

Like other whitefly-transmitted geminiviruses, STLCV is transmitted in a persistent, circulative manner by the whitefly *Bemisia tabaci,* and increasingly longer exposure to virus-infected plants results in increasingly greater transmission frequency. Viruliferous whiteflies can transmit the virus for at least 10 days (or for their lifetimes) if they have had a sufficient acquisition access period. The virus is not transmitted through true seed.

Natural hosts of STLCV are pepper and tomato, either of which may possibly serve as the most important reservoir of the virus for subsequent infection of either crop. Experimental hosts include *Datura stramonium, Malva parviflora, N. benthamiana, Phaseolus vulgaris,* pepper, and tomato.

Disease Cycle and Epidemiology

The disease cycle for STLCV is not well studied, but the virus is prevalent in Mexico, and at least one isolate has been identified in Costa Rica. The virus is transmitted by indigenous biotypes and biotype B of *B. tabaci,* as are most other begomoviruses in the subtropical Americas.

STLCV was discovered in Mexico just prior to the documented establishment of *B. tabaci* biotype B in the region. However, eggplant has not been examined as a prospective host of the virus, although it is commonly cultivated near peppers and tomatoes in the region. Biotype A, which once predominated, only rarely colonizes eggplant, but biotype B is now common and readily colonizes eggplant, which raises the possibility that eggplant is a natural, asymptomatic reservoir of STLCV. It is also possible that the propensity of biotype B for eggplant has provided additional selection pressure favoring widespread natural infection of eggplant and greater dispersion of STLCV in recent years.

With the even more recent emergence of newly reported tomato-infecting begomoviruses associated with *B. tabaci* biotype B, it is not known how widespread or how prevalent STLCV is in vegetable production on the west coast of Mexico or elsewhere in the Americas. Another begomovirus isolate, designated Pepper virus field isolate C, caused STLCV-like symptoms in pepper in Weslaco, Texas, and its CP nucleotide sequence shared more than 95% identity with that of STLCV, indicating STLCV was present in pepper fields in Texas as early as 1988–89, just after *B. tabaci* biotype B invaded the Rio Grande Valley and northern Mexico.

Control

Whitefly control to reduce the vector population in the field, removal of infected plants early in the season, and weed control near pepper fields are practiced to manage STLCV. However, as with other whitefly-transmitted viruses, infected pepper and tomato plants serve as sources of the virus and the vector, and growers have little opportunity to prevent damage. Different pepper species and cultivars appear to vary in the degree of damage caused by STLCV infection. No resistant varieties are available.

Selected References

Brown, J. K. 2001. Molecular markers for the identification and global tracking of whitefly vector–begomovirus complexes. Virus Res. 71:233–260.

Brown, J. K., Idris, A M., and Fletcher, D. C. 1993. Sinaloa tomato leaf curl virus, a newly described geminivirus of tomato and pepper in west coastal Mexico. Plant Dis. 77:1262.

Brown, J. K., Idris, A. M., Torres-Jerez, I., Banks, G. K., and Wyatt, S. D. 2001. The core region of the coat protein gene is highly useful for establishing the provisional identification and classification of begomoviruses. Arch. Virol. 146:1–18.

Idris, A. M., and Brown, J. K. 1998. Sinaloa tomato leaf curl gemini-virus: Biological and molecular evidence for a new subgroup III virus. Phytopathology 88:648–657.

Idris, A. M., Rivas-Platero, G., Torres-Jerez, I., and Brown, J. K. 1999. First report of Sinaloa tomato leaf curl geminivirus in Costa Rica. Plant Dis. 83:303.

(Prepared by J. K. Brown)

Tobacco etch virus

Tobacco etch virus (TEV) infects primarily solanaceous crops in North and South America. It has been reported as far north as Canada, throughout the United States, and in Mexico, Puerto Rico, Jamaica, Venezuela, and Sudan. TEV is one of the most damaging viruses affecting pepper in the United States. Disease incidence can be as high as 100%, with yield reductions of up to 70%. Pepper, tobacco, and tomato are the most important economic hosts of TEV in nature. However, over 120 species in 19 dicotyledonous families are susceptible to several strains of the virus.

Symptoms

Symptoms of TEV-infected pepper plants include mottle or mosaic (Plate 66), leaf distortion, and general stunting. The plants may produce fruit with severe mosaic (Plate 67). These symptoms can be confused with those caused by other viruses, including *Potato virus Y* (PVY) and *Pepper mottle virus* (PepMoV). Root necrosis and severe wilting develop in Tabasco pepper plants infected with TEV, followed by the death of the plants. In TEV epidemics in commercial pepper crops, a steady increase in the incidence of the virus over the entire growing season is usually observed. Pepper plants infected at a young age are usually more severely stunted than plants infected at a more mature stage of growth. Plants infected at an early stage also tend to produce small, malformed fruit, so that their yield is greatly reduced, compared to that of plants infected at a later stage.

Diagnostic hosts of TEV include *Datura stramonium*, *Nicotiana tabacum*, and *Capsicum frutescens* cv. Tabasco. Local lesion hosts include *Chenopodium* spp. and *Physalis peruviana*. TEV can be differentiated from PVY and PepMoV by biological assays: TEV, but not PVY, infects *D. stramonium;* PepMoV, but not TEV, induces local lesions on inoculated leaves of Tabasco pepper plants; and TEV infection results in the formation of nuclear inclusions, which do not occur in infection by PVY or PepMoV. These viruses can also be distinguished from one another by serological tests.

Causal Agent

TEV is a member of the genus *Potyvirus,* in the family *Potyviridae.* The virions are flexuous rods, 730 × 12–13 nm, containing single-stranded RNA that is translated into a single polyprotein. Individual TEV proteins are autocatalytically cleaved from the polyprotein by virus-encoded proteases that reside within the polyprotein.

Several strains or variants of TEV have been reported. Three strains were identified in Florida; five strains were reported in California, differentiated by the reactions of several resistant pepper cultivars. Few, if any, serological differences have been observed among isolates. TEV induces cytoplasmic cylindrical inclusions in host cells as well as crystalline nuclear inclusions.

Disease Cycle and Epidemiology

Natural epidemics of TEV infection in commercial pepper fields have been reported in California, Florida, Georgia, and Illinois. Weeds such as horse-nettle (*Solanum carolinense*), thistle (*Cirsium vulgare*), lamb's-quarters (*Chenopodium album*), jimson-weed (*D. stramonium*), and *Physalis* spp. have been reported as TEV reservoirs. Other solanaceous crops, such as tobacco and tomato, may also serve as sources of the virus.

Primary and secondary spread of TEV occurs by nonpersistent transmission by aphids. Acquisition and inoculation probes of 10 sec are sufficient for transmission, and the aphids can remain viruliferous for 1 to 4 hr. More than 10 species of aphids have been identified as vectors of TEV, including *Myzus persicae, Macrosiphum euphorbiae,* and *Aphis fabae.* No seed transmission of TEV has been reported in any host.

Control

Host resistance seems to be the best means of reducing losses due to TEV in commercial pepper production. Several sources of resistance have been reported. Some cultivars, including Agronomico 10C-5, Avelar, Delray Bell, VR4, Jalaro, and PI 152225, are apparently resistant to many of the virus isolates. Several resistant bell pepper cultivars are available from commercial seed companies. Horticultural mineral oil sprays (e.g., JMS Stylet-Oil), sticky traps, and reflective aluminum mulch have also been reported to provide limited control.

Selected References

Abdalla, O. A., Desjardins, P. R., and Dodds, J. A. 1991. Identification, disease incidence, and distribution of viruses infecting peppers in California. Plant Dis. 75:1019–1023.

Ariyaratne, I., Hobbs, H. A., Valverde, R. A., Black, L. L., and Dufresne, D. J. 1996. Resistance of *Capsicum* spp. genotypes to tobacco etch potyvirus isolates from the Western Hemisphere. Plant Dis. 80:1257–1261.

Benner, C. P., Kuhn, C. W., Demski, J. W., Dobson, J. W., Colditz, P., and Nutter, F. W., Jr. 1985. Identification and incidence of pepper viruses in northeastern Georgia. Plant Dis. 69:999–1001.

Black, L. L., Green, S. K., Hartman, G. L., and Poulos, J. M. 1991. Pepper diseases: A field guide. Asian Veg. Res. Dev. Cent. Publ. 91-347.

Demski, J. W. 1979. The epidemiology of tobacco etch virus-infected *Cassia obtusifolia* in relation to pepper. Plant Dis. Rep. 63:647–650.

Kuhn, C. W., Nutter, F. W., Jr., and Padgett, G. B. 1989. Multiple levels of resistance to tobacco etch virus in pepper. Phytopathology 79:814–818.

Makkouk, K. M., and Gumpf, D. J. 1974. Further identification of naturally occurring virus diseases of pepper in California. Plant Dis. Rep. 58:1002–1006.

Purcifull, D. E., and Hebert, E. 1982. Tobacco etch virus. Descriptions of Plant Viruses, no. 258. Commonwealth Mycological Institute and Association of Applied Biologists, Kew, England.

Sowell, G., Jr., and Demski, J. W. 1977. Resistance of plant introductions of pepper to tobacco etch virus. Plant Dis. Rep. 61:146–148.

Weinbaum, Z., and Milbrath, G. M. 1976. The isolation of tobacco etch virus from bell peppers and weeds in southern Illinois. Plant Dis. Rep. 60:469–471.

Zitter, T. A. 1972. Naturally occurring pepper virus strains in south Florida. Plant Dis. Rep. 56:586–590.

(Prepared by B. B. Reddick)

Tobacco mosaic virus and Tomato mosaic virus

Tobacco mosaic virus (TMV) and *Tomato mosaic virus* (ToMV) infect species and cultivars of the genus *Capsicum* wherever the crop is grown. They also infect tomato and tobacco. Yield losses in pepper due to TMV and ToMV infection can be substantial, ranging from 30 to 70% when resistant cultivars are not available. The viruses can spread extensively when peppers are field-grown from transplants or are handled frequently.

Symptoms

Symptoms caused by TMV and ToMV are generally similar in appearance but can vary greatly with the temperature, light intensity, day length, age of the plant when it was infected, strain of the virus, and cultivar. Chlorotic mosaic and distortion of leaves are common symptoms (Plate 68), and systemic necrosis and defoliation sometimes occur. In resistant cultivars, necrotic local lesions may develop on leaves inoculated with TMV or ToMV (Plate 69). Plants infected as seedlings or during early stages of growth may be stunted. Affected plants produce disfigured fruit, which is usually small and has distinct chlorotic or necrotic areas.

Causal Agents

TMV and ToMV are members of the genus *Tobamovirus.* They are rod-shaped particles, approximately 300 × 18 nm, each containing a single strand of single-stranded genomic RNA. The virus particles are very stable. Both viruses are easily sap-transmitted and have no known insect vectors.

Pepper mild mottle virus (PMMV) was once considered one of the many pepper-infecting strains of TMV but is now classified as a separate species of *Tobamovirus.* All three viruses are serologically distinct in double-antibody sandwich enzyme-linked immunosorbent assay (ELISA) with some common bioassay and differential host ranges. Necrotic local lesions occur on *Datura stramonium* and *Nicotiana tabacum* cvs. Turkish, Turkish Samsun, Samsun, White Burley, Burley, and Xanthi, all of which can be considered diagnostic hosts of tobamoviruses. Cross-reactivity between the species of tobamoviruses can occur in indirect ELISA.

Disease Cycle and Epidemiology

TMV- and ToMV-infected seed, fruit, leaves, stems, and root debris are common sources of inoculum for tobamovirus infections. In susceptible pepper cultivars, the viruses are distributed throughout the plant. They can also be present on and under the seed coat and in the endosperm. TMV and ToMV can be spread from plant to plant on hands, tools, trays, pots, stakes, twine, and clothing and in pollination, pruning, and other hands-on cultural practices.

Because of the stability of these viruses, infectious particles can contaminate plant surfaces, resulting in further spread of the disease. Both viruses can remain viable for several years in plant debris in dry soil. Particles rapidly lose their infectivity when debris remains in moist soil. Infectious ToMV has also been detected in fog.

TMV and ToMV have very wide host ranges, infecting over 200 plant species. Members of the Solanaceae, including sweet pepper, tomato, and tobacco, are susceptible. Many hot peppers are immune or have hypersensitive resistance to TMV and ToMV.

Control

Enforcement of strict sanitation in greenhouse pepper production and during harvest is essential to prevent infection and to minimize the spread of TMV and ToMV. Cultural practices and sanitation are important, such as restricting access to a crop, washing hands and equipment with a soap solution after contact with a plant before touching the next plant or row, and washing hands and equipment before entering and after leaving a greenhouse. Roguing symptomatic plants plus all adjacent plants and rotation to a nonsolanaceous crop also help to minimize disease incidence and spread. There are also many accounts of successful prevention of tobamovirus spread by coating hands, plants, and equipment in a solution of powdered instant nonfat milk.

The use of seed that has been tested and treated (indexed) for tobamoviruses is also recommended. However, indexed pepper seed is not guaranteed to be free of TMV and ToMV. A 2-hr seed treatment with a 10% solution of trisodium phos-phate (Na_3PO_4) is important in the management of these viruses, but it is not completely effective. Virus particles under the seed coat and in the endosperm are not inactivated by the treatment. Tobamoviruses are not a significant problem in open-field pepper production when indexed seed is planted and precautionary sanitation practices are established.

The use of resistant cultivars can reduce the spread of infection from a point of origin. Resistance to ToMV and TMV in peppers is controlled by four alleles at the L locus. The alleles are derived from various species of *Capsicum* and initiate a localized hypersensitive reaction (Plate 69). Plants containing the L_1 allele are resistant to almost all strains of ToMV and TMV. Some strains of PMMV can overcome the resistance conferred by L_1 but not that conferred by L_3 and L_4. Varieties containing L_3 and L_4 are used in protected culture areas where tobamoviruses are present.

Selected References

Beczner, L., Rochon, D. M., and Hamilton, R. I. 1997. Characterization of an isolate of Pepper mild mottle tobamovirus occurring in Canada. Can. J. Plant Pathol. 19:83–88.

Castello, J. D., Lakshman, D. K., Tavantzis, S. M., Rogers, S. O., Bachand, G. D., Jagels, R., Carlisle, J., and Liu, Y. 1995. Detection of infectious tomato mosaic tobamovirus in fog and clouds. Phytopathology 85:1409–1412.

Hollings, M., and Huttinga, H. 1976. Tomato mosaic virus. Descriptions of Plant Viruses, no. 156. Commonwealth Mycological Institute and Association of Applied Biologists, Kew, England.

Tobias, I., Rast, A. T. B., and Maat, D. Z. 1982. Tobamoviruses of pepper, eggplant, and tobacco: Comparative host reactions and serological relationships. Neth. J. Plant Pathol. 88:257–268.

Villálon, B. 1981. Breeding peppers to resist virus diseases. Plant Dis. 65:557–562.

Watterson, J. C. 1993. Development and breeding of resistance to pepper and tomato viruses. Pages 80–101 in: Resistance to Viral Diseases in Vegetables—Genetics and Breeding. M. M. Kyle, ed. Timber Press, Portland, Ore.

Wetter, C. 1984. Serological identification of four tobamoviruses infecting pepper. Plant Dis. 68:597–599.

Wetter, C., and Conti, M. 1988. Pepper mild mottle virus. Descriptions of Plant Viruses, no. 330. Commonwealth Mycological Institute and Association of Applied Biologists, Kew, England.

Zaitlin, M., and Israel, H. W. 1975. Tobacco mosaic virus (type strain). Descriptions of Plant Viruses, no. 151. Commonwealth Mycological Institute and Association of Applied Biologists, Kew, England.

Zitter, T. A. 1991. Tomato mosaic and tobacco mosaic. Page 39 in: Compendium of Tomato Diseases. J. B. Jones, J. P. Jones, R. E. Stall, and T. A. Zitter, eds. American Phytopathological Society, St. Paul, Minn.

(Prepared by P. T. Himmel)

Tomato spotted wilt virus

Tomato spotted wilt virus (TSWV) infects peppers and a wide range of other crops across temperate, subtropical, and tropical regions throughout the world. Spotted wilt was first reported in 1915, affecting tomatoes in Australia, and was later demonstrated to be of viral origin. In addition to pepper and tomato, major crops susceptible to TSWV infection include chrysanthemum, lettuce, potato, peanut, and tobacco. Infection rates of 50–90% lead to major losses in commercial vegetable crops, making TSWV one of the most economically destructive plant viruses of recent times.

Symptoms

The appearance and severity of spotted wilt symptoms in pepper vary widely, depending on the cultivar, virus isolate, stage of plant growth at the time of infection, and environmental conditions. Plants infected at the transplant stage are usually severely stunted throughout the growing season (Plate 70) and frequently yield no fruit. Plants infected later in the season may exhibit chlorotic or necrotic flecking (Plate 71) or necrotic ring spots (Plate 72) on leaves and stems. Flower and leaf drop occur in some cultivars. Chlorotic or necrotic spots, ring patterns, or mosaic may develop on the fruit of infected plants; these symptoms are especially evident as an undesirable fruit color at maturity (Plate 73).

Causal Agent

TSWV is the type member of the plant-infecting genus *Tospovirus*, in the family *Bunyaviridae*, a large group of predominantly vertebrate- and insect-infecting RNA viruses. TSWV forms pleomorphic particles, 80–120 nm in diameter, with surface projections composed of two viral glycoproteins (G1 and G2). The TSWV genome consists of one negative and two ambisense single-stranded RNAs that adopt a pseudocircular or panhandle conformation. Each genomic RNA is encapsidated by multiple copies of the viral nucleocapsid (N) protein to form ribonucleoprotein structures, also known as nucleocapsids. The nucleocapsids are enveloped in a host-derived membrane bilayer along with the viral L protein, the putative RNA-dependent RNA polymerase.

More than 800 plant species, both dicots and monocots, in more than 80 plant families are susceptible to TSWV. The families Solanaceae and Asteraceae contain the largest numbers of susceptible species.

Disease Cycle and Epidemiology

TSWV can be transmitted mechanically in the laboratory but in the field it is transmitted from plant to plant almost exclusively by several species of thrips (order Thysanoptera, family Thripidae). The western flower thrips (*Frankliniella occidentalis*) and the tobacco thrips (*F. fusca*) are major vectors, and *F. schultzei, F. intonsa, F. bispinosa, Thrips tabaci,* and *T. setosus* can also transmit the virus. Only larval thrips can acquire TSWV, while both larval and adult thrips can transmit it in a persistent though often sporadic fashion. TSWV replicates in the thrips as well as in its plant host. The virus and its vectors are frequently spread in the transport of ornamentals and vegetable transplants. Adjoining fields of agronomic crops (often peanut or tobacco) and nearby weeds are also common sources of both the virus and vectors.

Control

The extremely wide and overlapping host range of TSWV and its vectors makes control difficult. Adding to the difficulty is the scarcity of host plant resistance genes and the large number of weed and ornamental hosts providing between-crop virus reservoirs. Multicomponent management approaches are most effective. The use of virus-free transplants is essential. Thrips-proof screens can prevent or delay infection in greenhouse production. Chemical control of the vectors is generally not recommended, as it is difficult to achieve effective coverage with chemicals to kill thrips before they are able to transmit the virus. There are few, if any, commercial pepper cultivars with TSWV resistance, although there are promising developments in pathogen-derived (transgenic) resistance in tomato and tobacco.

Selected References

Adkins, S. 2000. Tomato spotted wilt virus: Positive steps towards negative success. Mol. Plant Pathol. 1:151–157.

Cho, J. J., Mau, R. F. L., Gonsalves, D., and Mitchell, W. C. 1986. Reservoir weed hosts of tomato spotted wilt virus. Plant Dis. 70: 1014–1017.

German, T. L., Ullman, D. E., and Moyer, J. W. 1992. *Tospoviruses:* Diagnosis, molecular biology, phylogeny, and vector relationships. Annu. Rev. Phytopathol. 30:315–348.

Goldbach, R., and Peters, D. 1994. Possible causes of the emergence of tospovirus diseases. Semin. Virol. 5:113–120.

Sherwood, J. L., German, T. L., Whitfield, A. E., Moyer, J. W., and Ullman, D. E. 2000. Tomato spotted wilt. Pages 1030–1031 in: Encyclopedia of Plant Pathology. O. C. Maloy and T. D. Murray, eds. John Wiley & Sons, New York.

(Prepared by S. Adkins)

Postharvest Diseases and Disorders

Postharvest Losses

Losses of pepper fruit after harvest usually vary with the weather in the production field. Warm, rainy weather leads to more postharvest decay in packed peppers than any other environmental conditions.

A survey in greater New York City in 1974 recorded losses of 7.1% of Florida-grown peppers in wholesale markets and 9.2% in retail markets. Thus, more than 16% of packed fruit was lost before it could be marketed. Consumer losses were estimated to be 1.4%, on the basis of samples from retail markets. Parasitic diseases were the primary cause of losses in wholesale markets, affecting 4% of the fruit examined, whereas in supermarkets the greatest losses (5.0%) were due to physical injury. In samples purchased in retail sales, 7.2% of the fruit was damaged by parasitic diseases. One of nine shipments of peppers examined at wholesale markets contained more than 10% decayed fruit.

In an earlier survey of rail shipments unloaded at New York City, decay per car lot averaged 3.6% over a seven-year period, with half the damage due to bacterial soft rot. Other diseases were identified as Rhizopus rot, gray mold rot (Botrytis fruit rot), and watery soft rot, and some infections were categorized as "other rots."

Fruit shipped from Florida to Rotterdam was examined for decay between January and April 1979. The average incidence of decay was over 42% after arrival and simulated wholesale and retail handling. Bacterial soft rot, with 18.9% incidence, was the most common disease.

Alternaria rot was found in 16.2% of peppers in retail markets in another study. Shipment of the peppers at 6°C, in which some chilling injury may have occurred, is likely to have contributed to the development of the disease. According to that report, mature green bell peppers should not be stored at temperatures below 7°C. Chilling injury occurs at lower temperatures and predisposes peppers to Alternaria rot and sheet pitting.

Postharvest Diseases

Postharvest decays of pepper fruit include, in descending order of importance, bacterial soft rot (caused by *Erwinia carotovora* subsp. *carotovora*), Alternaria rot (*Alternaria* spp.); Botrytis fruit rot (*Botrytis cinerea*), Rhizopus rot (*Rhizopus stolonifer*) and sour rot (*Geotrichum candidum*). Various other diseases also affect harvested peppers, particularly if the fruit is ripe or injured by chilling.

Anthracnose (caused by *Colletotrichum capsici* and *C. gloeosporioides*) and Phytophthora blight (*Phytophthora capsici* and

P. nicotianae var. *parasitica*) can develop in fruit with incipient lesions or latent infections at harvest, but these diseases have not been significant in wholesale or retail markets.

The postharvest diseases most likely to cause losses are bacterial soft rot and Rhizopus rot, which can spread in packages or boxes of pepper fruit, creating nests of decay. Bacterial soft rot develops more rapidly than the other decays and has been more commonly found in market surveys and thus should be considered the most important postharvest disease.

Selected References

Cantwell, M. 2000. Bell pepper. Produce Facts: Recommendations for Maintaining Postharvest Quality. On-line publication. Postharvest Technology Research and Information Center, Department of Pomology, University of California, Davis.

Ceponis, M. J., and Butterfield, J. E. 1974. Causes of cullage of Florida bell peppers in New York wholesale and retail markets. Plant Dis. Rep. 58:367–369.

Ceponis, M. J., and Butterfield, J. E. 1974. Market losses in Florida cucumbers and bell peppers in metropolitan New York. Plant Dis. Rep. 58:558–560.

McColloch, L. P. 1966. Chilling injury and Alternaria rot of bell peppers. U.S. Dep. Agric. Mark. Res. Rep. 536.

McDonald, R. E., and de Wildt, P. P. Q. 1980. Cause and extent of cullage of Florida bell peppers in the Rotterdam terminal market. Plant Dis. 64:771–772.

Wiant, J. S., and Bratley, C. O. 1948. Spoilage of fresh fruits and vegetables in rail shipments unloaded at New York City 1935–42. U.S. Dep. Agric. Circ. 773.

(Prepared by J. A. Bartz)

Bacterial Soft Rot

A bacterial soft rot of peppers was first reported in 1901. Both ripe and green fruit rotted rapidly after inoculation with a bacterium now called *Erwinia carotovora* subsp. *carotovora,* and green fruit appeared somewhat more susceptible than red fruit. Bacteria belonging to at least five different genera macerate pepper fruit and could be considered causes of bacterial soft rot. The disease can involve the stem as well as various parts of the fruit.

Symptoms

Lesions begin as sunken, light- to dark-colored, water-soaked areas in or around the edge of wounds on the fruit, peduncle, or lobular calyx. Typically, chlorophyll is not immediately destroyed, and the lesions do not have a halo or appear necrotic. The water-soaked areas expand rapidly. Affected tissues quickly lose texture. Lesions may form on the inner surfaces of fruit that has been infiltrated with water from rainfall, water handling, or hydrocooling. Lesion contents may disperse to adjacent and nearby fruits, causing secondary infections and nests of decay. Bacterial ooze may develop at infection sites or on enlarging lesions, and secondary fungi may invade rotted tissue (Plate 74).

Causal Organisms

Strains of *Erwinia carotovora* subsp. *carotovora* (Jones) Bergey et al. (syns. *Bacillus carotovorus* Jones, *E. aroideae* (Town.) Holland, and *Pectobacterium carotovorum* (Jones) Waldee), *E. carotovora* subsp. *atroseptica* (van Hall) Dye, and *E. chrysanthemi* Burkholder et al. have been isolated from bacterial soft rot lesions on pepper fruit and shown to be capable of initiating the disease in inoculated fruit. The relative aggressiveness of these three bacteria is related to temperature.

At 23°C, *E. carotovora* subsp. *carotovora* and *E. chrysathemi* caused nearly twice as much decay as *E. carotovora* subsp. *atroseptica*. In contrast, at 10°C, *E. carotovora* subsp. *atroseptica* was twice as aggressive as *E. carotovora* subsp. *carotovora* and more than five times as aggressive as *E. chrysanthemi*.

The pectolytic strains of *Erwinia* are facultatively anaerobic, strongly pectolytic, Gram-negative rods (0.5–1.0 × 1.0–2.0 µm), which are mobile by means of peritrichous flagella. Cells of these bacteria are found singly, in pairs, or occasionally in short chains. They do not form spores. Colonies on nutrient agar are grayish white, slightly raised, smooth, glistening, round, and translucent. A single cell can multiply to form a colony on nutrient agar within 18 hr at 22°C; the optimal growth temperature is 27–30°C. Most strains of *E. carotovora* subsp. *carotovora* and *E. chrysanthemi* will grow at temperatures above 36°C, but few strains of *E. carotovora* subsp. *atroseptica* will do so. The minimum temperature for growth was reported to be 3°C, with inconsistent indications that *E. carotovora* subsp. *atrospetica* has a lower temperature range than the other two. These pectolytic strains produce deep pits on crystal violet polypectate medium. Some strains produce initial pitting within 8 hr of streaking the surface of the medium.

Strains of pectolytic fluorescent *Pseudomonas* spp. (Treisan Migula biovar II, syn. *P. marginalis* (Brown) Stevens) can also cause a decay of pepper fruit. The fluorescent pseudomonads are psychotropic, and therefore they are likely to have lower optimum temperatures than the *Erwinia* spp. The pectolytic strains of *Pseudomonas* spp. form cells measuring 0.7–0.8 × 2.0–2.8 µm. On King's medium B, they produce green pigments that fluoresce under UV light. They do not ferment carbohydrates and consequently are not facultative anaerobes. These strains have been isolated from water-soaked areas on pepper fruit but are not considered aggressive enough to cause major losses of fruit.

Pectolytic strains of *Bacillus* spp., *Xanthomonas campestris* (Pammel) Dowson, and *Cytophaga* spp. have also been associated with soft rots of pepper fruit, but their aggressiveness relative to that of soft rot strains of *Erwinia* spp. is unknown.

Disease Cycle and Epidemiology

Bacteria causing soft rot of pepper fruit are ubiquitous and present in all production areas. Strains of soft rot *Erwinia* spp. do not survive well in fallowed soil. They are predominately associated with surface water and with the rhizosphere of certain plants. In contrast, the pectolytic *Pseudomonas* spp. are present in soil, particularly under crop canopies. Rain and fog favor the rapid development of bacterial populations on plants. Warm, moist weather is considered highly favorable for infection.

None of the soft rot bacteria can directly penetrate intact fruit surfaces. However, the fleshy peduncle and associated lobed calyx are highly susceptible and are frequently the point of initial attack, particularly on bell peppers. Moist conditions in the field from rainfall or irrigation appear to increase the succulence of the peduncle and hence its susceptibility to bacterial soft rot. Nitrogen fertilization has also been associated with increased susceptibility.

Most bell pepper cultivars lack a defined abscission zone, and the peduncle may be broken as the fruit is snapped from the plant at harvest. The broken peduncle can quickly become infected if the fruit is held under warm, moist conditions, especially if free water is present. The soft rot progresses from the broken surface into the calyx and then into the fruit. Pungent varieties often have an abscission zone, and they do not suffer stem infections unless damaged during harvest.

Soft rot infection can also begin in the field prior to harvest. Fruit injured by sunscald or insect feeding can become infected by soft rot bacteria. Openings in the fruit surface as a result of sun injury allow water from rainfall or irrigation to enter the fruit. The inner surfaces of the fruit lack sufficient wax to protect it from direct attack by soft rot bacteria. Inoculation can also occur if storm-driven rain enters the fruit. Decay in the field results in macerated tissues and inoculum that disperses to nearby fruit on the plant. Field crews can also disperse inoculum from diseased to healthy fruit during harvest.

Pepper fruit is usually hand-picked, placed in harvest containers, and then packed at a central facility. Contamination of harvest containers by soft rot bacteria increases the incidence of disease. Equipment in the packing line can also disperse bacteria to infection courts. Peppers are either dry- or wet-dumped (containers are emptied either onto a dry, padded surface or into a water bath). Dry dumps are difficult to sanitize and can cause excessive abrasion of fruit, particularly if the crop has been grown in coarse-textured soil and splashing rain has deposited soil on the surface of the fruit. Wet dumps can be continuously sanitized by chlorination. However, if the fruit cools or is forced deep under the surface of the bath, water may enter it, carrying decay pathogens or other hazardous microorganisms. Additionally, residual water on fruit may be difficult to remove. If fruit is wet when packed or packaged, postharvest decays are likely to develop.

The method by which peppers are cleaned also affects the postharvest incidence of bacterial soft rot. Dry brushing followed by waxing can inoculate the fruit, particularly if there is a high incidence of decay in the field. Water washes must contain a sanitizer, and the fruit must be dried after washing.

The jumble pack for bell peppers and the film-overwrapped tray for most pepper types also increase the likelihood that bacterial soft rot will develop. In a jumble pack, random distribution of the fruit and, frequently, overfilling can injure the fruit and restrict air movement inside the package. The injuries create infection courts for soft rot bacteria, and the stagnant air allows humidity to rise and retards the drying of surfaces, thus promoting the onset of the disease. The film-overwrapped tray prevents moisture loss and associated shrivel, but if condensation develops within the package, various decays, including bacterial soft rot, are likely to occur. Areas where the film is stretched over the fruit are prone to attack.

Control

Four general strategies are used to control bacterial soft rot of pepper fruit: avoiding conditions in the field that favor the development of the disease, maintaining sanitary conditions during and after harvest, cooling fruit promptly after harvest, and packing fruit in vented packages held at uniform storage temperatures to avoid condensation on the fruit. These strategies are mainly preventative, although certain cooling methods are at least partially therapeutic.

In the field, chewing insects must be controlled, since each feeding site is a potential infection court for soft rot bacteria. Pepper crops should be planted so that the anticipated harvest date does not occur during a rainy season. Nitrogen fertilization should follow cooperative extension recommendations, and nitrogen should not be applied at an excessive rate. Irrigation should complement rainfall and not subject plants to excessive wetness.

Harvesting or other manipulations of plants should not be conducted when the foliage is wet or cold, as excessive injury and dispersal of microorganisms is likely under those conditions. The harvest containers should be cleaned and sanitized before and at least daily during the harvest. All surfaces of the containers that come into contact with fruit should be smooth. Smooth, clean surfaces can be sanitized with chlorinated water or other approved chemicals. To control all decay organisms, surfaces should be flooded for at least 2 min with a chlorine solution with a free chlorine concentration of 200 ppm at pH 6–7.5. Field crews should be trained to discard partially decayed fruit and to avoid unnecessary fruit injury. Dry dumps and brush waxer units at packinghouses should be periodically

cleaned and sanitized. The dry brushes should be treated with an application of chlorine at a concentration of 600 ppm and then allowed to stand for 30 min, and the treatment should be repeated several times a day. In wet dumps and spray washers, the free chlorine concentration should at all times be at least 100 ppm at pH 6.0–7.5. If fungal decay is observed, the chlorine concentration should be increased to 200 ppm.

Free water should be removed from fruit surfaces by a combination of air drying and cooling. Jumble packs should not be overfilled and should have proper venting. Film-overwrapped packages should be perforated to prevent condensation.

Packaged fruit should be promptly cooled to 7.5 to 10°C by forced air. In forced-air cooling, cooled air is pulled through stacks of packages, not just moved around them. This type of cooling system will remove some of the moisture from peduncles and calyxes, making them less susceptible to bacterial soft rot. Hydrocooling is not recommended for peppers, because it tends to add moisture to the fruit rather than remove it. Air for forced-air cooling can be cooled either mechanically or by evaporative methods. Evaporative coolers are more effective in less humid growing areas, such as California or Arizona. Vacuum cooling can be used, particularly if the peduncles have started to break down. The ratio of surface area to volume of peppers does not allow efficient vacuum cooling, but this method removes water from the peduncle and reportedly can arrest lesions that have started to form. Cooling must be completed before the packaged fruit is loaded into trucks, because the cooling units on trucks are designed to maintain load temperatures and not reduce them. The cold chain must be carefully maintained to avoid condensation and chilling injury.

Selected References

Burr, T. J., and Schroth, M. N. 1977. Occurrence of soft-rot *Erwinia* spp. in soil and plant material. Phytopathology 67:1382–1387.

Coplin, D. L. 1980. *Erwinia carotovora* var. *carotovora* on bell peppers in Ohio. Plant Dis. 64:191–194.

Harrison, M. D., Franc, G. D., Maddox, D. A., Michaud, J. E., and McCarter-Zorner, N. J. 1987. Presence of *Erwinia carotovora* in surface water in North America. J. Appl. Bacteriol. 62:565–570.

Johnson, H. B. 1964. Effect of hot water treatments and hydrocooling on bacterial soft rot in bell peppers. U.S. Dep. Agric., Agric. Mark. Serv., Mark. Qual. Res. Div. No. 517.

Johnson, H. B. 1966. Bacterial soft rot in bell peppers: Cause and commercial control. U.S. Dep. Agric. Mark. Res. Rep. 738.

McColloch, L. P., Cook, H. T., and Wright, W. R. 1968. Market diseases of tomatoes, peppers, and eggplants. U.S. Dep. Agric. Agric. Handb. 28.

Stommel, J. R., Goth, R. W., Haynes, K. G., and Kim, S. H. 1996. Pepper (*Capsicum annuum*) soft rot caused by *Erwinia carotovora* subsp. *atroseptica*. Plant Dis. 80:1109–1112.

(Prepared by J. A. Bartz)

Alternaria Rot

Alternaria rot occurs in all production areas, affecting pepper fruit having a predisposing injury. The causal agent is a weak pathogen and normally does not spread in harvested fruit. Fruit still attached to the plant may have an internal rot, which is, at least initially, limited to the seeds and placenta. Such fruit appears normal and would not be culled at the packinghouse. Internal mold has been found in bell peppers grown in Israel, chili and bell peppers grown in the western United States, and red bell peppers grown in Florida. Surface lesions may form at growth cracks, at the site of wounds received during harvest, or on tissue that has been damaged by chilling, sunscald, calcium deficiency, insect injury, or heat.

Symptoms

Small, circular, slightly sunken spots may develop anywhere on the fruit surface but often start at cracks or other surface injuries. Recently formed spots are usually similar in color to the adjacent healthy tissue, perhaps with grayish tones. As they enlarge, the spots develop into sharply sunken lesions. Damaged or ripening tissues appear necessary for lesion development, but some spots seem to develop in the absence of visible wounds. The spots eventually become covered with grayish to black mold (Plate 75). Tissues around the surface mold appear green to black. Copious spore production gives the surface a dusty appearance. In film-wrapped packages, the peduncle and calyx may become moldy. In packaged ripe fruit, mold may develop on the blossom scar, and when the fruit is cut open, the seeds and placenta will be found to be covered with mold.

Causal Organism

The cause of Alternaria rot of peppers is not completely clear, but the pathogen is most likely one or more small-spored *Alternaria* spp. The name commonly applied to the fungus found on peppers, tomatoes, and certain other fruit vegetables, *A. alternata* (Fr.:Fr.) Keissl., appears to be invalid. The fungus is a weak pathogen that has been reported to cause lesions in a wide range of harvested fruits and vegetables. It exists naturally as a saprophyte and is one of the first colonizers of dying plants. The conidia are pigmented black to dark olive and are dilute ovoid, obclavate, obpyriform, or, rarely, simple ellipsoidal. Each usually has a visible basal pore. All but the ellipsoidal form of conidia have a beak or apical cell that is shorter than the rest of the spore. The conidia may be muriform (containing both longitudinal and cross septa) and are usually formed in long chains.

Disease Cycle and Epidemiology

Primary inoculum is airborne and often arises from mold on debris of a wide array of plants. Initial infections may start on the stigma of the flower, leading to internal mold; on wounds or damaged tissues; or on drying stems and calyxes. Green fruit is usually not attacked unless it has been weakened by cold, calcium deficiency, or high temperature. As the fruit ripens, it becomes more susceptible. In one study, wound-inoculated green New Mexico chili peppers appeared quite resistant to rot, whereas lesions 1.7 to 2.2 cm in diameter developed within 10 days on fruit that was 10 to 100% red. However, Alternaria rot of green peppers has also been reported following long storage at moderate temperatures and in peppers with severe wounds or crushed tissues. The disease progresses most rapidly at 24 to 28°C. Refrigeration greatly slows disease development. In one study, lesions did not form on green fruit stored at 0°C but developed rapidly when the fruit was warmed to 18°C. Alternaria rot does not readily spread in stored fruit, but spores released from early infections can cause limited secondary spread.

Control

Alternaria rot is controlled by avoiding conditions that predispose peppers to infection both in the field and after harvest. There are no approved fungicides or biocontrol agents for postharvest application to pepper fruit. Planting should be scheduled so that the anticipated harvest date does not occur during very hot or very cold weather or when moisture is likely to be insufficient. Cultivars should be selected for providing good shading of developing fruit. The fertilization program should ensure adequate calcium. Rough handling during harvest, including the use of bulk packs and overloading of bins, should be avoided.

The fruit should be cleaned, packed, and then promptly cooled to 7.5–10.0°C. The washer, waxer, and other packing line equipment should be periodically cleaned and sanitized. Prepackaged fruit can be wrapped with film to prevent desicca-

tion of the peduncle and calyxes, but should not lead to condensation within the package. The storage room should be periodically cleaned. Pepper fruit in storage should be kept at high humidity to prevent the stems from drying excessively. A hotwater brush technique was recently described, in which peppers are brushed with water at 55°C for 12 sec, to control postharvest decay and maintain fruit quality during long-term shipment. Hot-water treatments have been reported to control *Alternaria* spp. on other fruit vegetables and to reduce the sensitivity of certain fruits to chilling injury.

Selected References

Fallik, E., Grinberg, S., Alkalai, S., Yekutieli, O., Wiseblum, A., Regev, R., Beres, H., and Bar-Lev, E. 1999. A unique rapid hotwater treatment to improve storage quality of sweet pepper. Postharvest Biol. Technol. 15:25–32.

Halfon-Meiri, A., and Rylski, I. 1983. Internal mold caused in sweet pepper by *Alternaria alternata:* Fungal ingress. Phytopathology 73: 67–70.

McColloch, L. P. 1962. Chilling injury and Alternaria rot of bell peppers. U.S. Dep. Agric. Mark. Res. Rep. 536.

McColloch, L. P., Cook, H. T., and Wright, W. R. 1968. Market diseases of tomatoes, peppers, and eggplants. U.S. Dep. Agric. Agric. Handb. 28.

Simmons, E. G., and Roberts, R. G. 1993. *Alternaria* themes and variations (73). Mycotaxon 48:109–140.

Wall, M. M., and Biles, C. L. 1993. Alternaria rot of ripening chile peppers. Phytopathology 83:324–328.

(Prepared by J. A. Bartz)

Botrytis Fruit Rot

Botrytis fruit rot, also known as gray mold rot, is a sporadic problem worldwide. It is associated with cool, moist conditions in production fields, chilling injury prior to harvest, and frost injury. It also occurs in greenhouse production. Proper handling and storage temperatures greatly diminish postharvest development of the disease in green peppers.

Symptoms

Small, cream-colored specks form anywhere on the fruit surface, often at wounds. They enlarge in a rhizoid pattern and develop into large, round, water-soaked lesions, in shades of gray mingled with the normal color of the fruit (Plates 76 and 77). Lesions generally do not develop on wound-free surfaces unless the fruit has been chilled or damaged by frost. Decayed tissues are softened and moist but have a firmer consistency than tissues affected by other fungal or bacterial soft rots. A brown to gray mold develops on the lesion surface under humid conditions. Under a hand lens or binocular dissecting microscope, grapelike clusters of conidia may be seen on the surface of the mold growth. Lesions may also form on the cut or torn end of the peduncle.

Causal Organism

Botrytis cinerea Pers.:Fr., the fungus causing Botrytis fruit rot, can be cultured on various media. Conidiophores are long and dichotomously branched at the tip. At the tip of each branch, conidia form on sterigmata in clusters resembling grapes (Fig. 5). The conidia are hyaline, single-celled, and ovoid, measuring 9.7–11.1 × 7.3–8.0 μm. Black to dark brown sclerotia may form on old lesions and readily form in old cultures. Petri dish cultures form lawns of conidia and no or few sclerotia if incubated continuously under fluorescent light. The sclerotia usually germinate directly to form mycelium, which then produces conidiophores and conidia. Rarely, sclerotia have been reported to produce apothecia and ascospores. Optimal temperatures for growth of the fungus are 18–23°C, and growth ceases at 32°C. However, the minimum temperature for growth is below 0°C. Botrytis fruit rot was reported to develop more rapidly in pepper at 10°C than at 13°C or above or at 7°C or below.

Disease Cycle and Epidemiology

Primary inoculum can be windborne from a very wide range of host plants. Saprophytic growth of *B. cinerea* on plant debris and germination of sclerotia in soil or on plant debris can also provide primary inoculum. *B. cinerea* is considered a weak pathogen, but it is adept at colonizing senescing leaves and flower petals of many plant species. It can penetrate plants directly by forming appressoria if a source of food is available. Leaves or fruit resting on the soil surface may be directly penetrated.

The disease does not progress to the fruit rot phase unless the fruit has been injured by cold. It is not clear whether rainfall or fog is required for fruit infection. Peppers are reported to become infected in cool weather, with or without high humidity, and a large canopy is an important predisposing factor. Low calcium nutrition in greenhouse culture appears to predispose pepper plants to infection. Tomato crops grown in soils low in calcium also have increased susceptibility to gray mold and Botrytis fruit rot.

Control

Avoidance of conditions that favor disease development is recommended. Crops that will be exposed to chilling in the field must be handled carefully. Injury to tissues from exposure to cold is a cumulative process, and therefore fruit from such crops should not be refrigerated or stored too long. The hot-water treatment recommended for decay control in general may have utility in controlling *B. cinerea*.

If gray mold occurs in the field, harvest containers and the packing line should be periodically sanitized. Spores from field infections on harvested fruit can contaminate surfaces and then spread to wounds on the fruit or the broken surface of the peduncle of healthy fruit. Chlorine concentrations for washers and

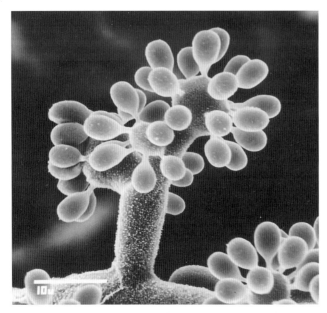

Fig. 5. Conidiophore and conidia of *Botrytis cinerea* (scanning electron micrograph). (Reprinted, by permission, from M. A. Ellis, R. H. Converse, R. N. Williams, and B. Williamson, eds., 1991, Compendium of Raspberry and Blackberry Diseases and Insects, American Phytopathological Society, St. Paul, Minn.)

wet dumps have not been established, but the degree of sensitivity of *B. cinerea* in general suggests that the free chlorine concentration should at all times be at least 100 ppm at pH 6.0–7.5.

Peppers in greenhouse culture must receive recommended levels of calcium. Greenhouse growers should also take steps to avoid high humidity and condensation on plants in cool weather.

Selected References

Brecht, J. K., Chen, W., Sargent, S. A., Cordasco, and Bartz, J. A. 1999. Exposure of green tomatoes to hot water affects ripening and reduces decay and chilling injury. Proc. Fla. State Hortic. Soc. 112: 138–143.

Elad, Y., Yunis, H., and Volpin, H. 1993. Effect of nutrition on susceptibility of cucumber, eggplant and pepper crops to *Botrytis cinerea*. Can. J. Bot. 71:602–608.

McColloch, L. P. 1966. Botrytis rot of bell peppers. U.S. Dep. Agric. Mark. Res. Rep. 754.

McColloch, L. P., Cook, H. T., and Wright, W. R. 1968. Market diseases of tomatoes, peppers, and eggplants. U.S. Dep. Agric. Agric. Handb. 28.

Stall, R. E. 1991. Gray mold. Pages 16–17 in: Compendium of Tomato Diseases. J. B. Jones, J. P. Jones, R. E. Stall, and T. A. Zitter, eds. American Phytopathological Society, St. Paul, Minn.

(Prepared by J. A. Bartz)

Rhizopus Rot

Rhizopus rot is usually a minor postharvest problem in peppers, because cool storage conditions inhibit the development of the pathogen. However, the disease has occurred in peppers in markets around the world and can cause losses during transit from production fields to markets.

Symptoms

Small, water-soaked spots form at the broken end of the peduncle, at wounds on the fruit, or on the inner walls of the fruit, and they quickly enlarge. Tissues in the lesions are soft but have some consistency, due to the ramification of the diseased tissues by coarse mycelium. Diseased fruit is soon engulfed by the pathogen. Under humid conditions, grayish white masses of mold structures (mycelium, sporangiophores, sporangia, and occasionally zygospores) develop over lesions. Clear liquid is released from diseased tissues. Nests of mold and decaying fruit may form in packages.

Causal Organism

Rhizopus stolonifer (Ehrenb.:Fr.) Vuill. (syn. *R. nigricans* Ehrenb.) is a ubiquitous zygomycete that produces sporangiospores in a globose, columellate sporangium. The mycelium is white to grayish. Sporangia change from white to black as they mature. They rupture easily, releasing many small spores. The mycelium is initially coenocytic (lacking cross-walls) and have many branches. Septa may eventually form. *R. stolonifer* grows rapidly on solid media and soon fills the air space in a petri dish. Spores may spill out around the edge of the dish cover, so that the organism can become a major contaminant in the laboratory.

Disease Cycle and Epidemiology

R. stolonifer is a widespread, common saprophyte. It has been reported to survive in dried deposits for up to 30 years. Contact between freshly harvested fruit and contaminated picking containers, storage bins, or packing line equipment is a primary means by which infection is initiated. High temperature and humidity and fresh wounds promote disease development. Optimum temperatures for the disease are in the range of 24–27°C, whereas little disease development occurs below about 10°C. Spores of the pathogen may be dispersed by wind or splashing rain. *R. stolonifer* is a wound pathogen, but in nests of decaying fruit it usually finds a way to penetrate the surfaces of adjacent fruit. If contaminated wash water enters pepper fruit, the pathogen can directly infect the inner walls.

Control

The key steps in controlling Rhizopus rot are to sanitize all equipment, maintain a chlorine concentration of at least 150 ppm at pH 6.0–7.5 in wet dumps and washers, prevent severe wounds, promptly dry wet surfaces of the fruit, and cool the fruit to below 10°C as quickly as possible.

Selected References

Baker, K. F. 1946. An epiphytotic of Rhizopus soft rot of green-wrap tomatoes in California. Plant Dis. Rep. 30:20–26.

Bartz, J. A. 1991. Rhizopus rot. Page 46 in: Compendium of Tomato Diseases. J. B. Jones, J. P. Jones, R. E. Stall, and T. A. Zitter, eds. American Phytopathological Society, St. Paul, Minn.

McColloch, L. P., Cook, H. T., and Wright, W. R. 1968. Market diseases of tomatoes, peppers, and eggplants. U.S. Dep. Agric. Agric. Handb. 28.

(Prepared by J. A. Bartz)

Chilling Injury

Chilling injury predisposes pepper fruit to decay, including bacterial soft rot. Fruit exposed to chilling in the field prior to harvest is less tolerant of refrigeration after harvest, which suggests that cold injury is cumulative. Safe minimum storage temperatures have been reported to be 7.2–7.5°C, whereas storage at 5°C for two weeks or longer leads to chilling injury. Freshly harvested peppers can be stored for three to five weeks at 7.5°C, if moisture loss is prevented. Ripe or ripening peppers are less sensitive to low temperatures than green peppers. Fruit injured by chilling may not show symptoms (sheet pitting

and surface scald) until after it has been allowed to warm to ambient temperature for marketing. However, most peppers stored at 0°C for 12 days or longer will lose their luster and become lifeless. Increased susceptibility to attack by decay pathogens is an initial symptom of chilling injury. In one study, bell peppers stored at 10°C were more prone to Botrytis fruit rot than peppers stored at 12.7°C.

Marine shipment of peppers requires prolonged storage. Peppers have been successfully stored up to 28 days after harvest when held at temperatures just above the level at which chilling injury occurs. However, in long-term refrigerated storage, water loss may occur, leading to shrivel. Prepackaging with a film overwrap prevents excessive moisture loss and extends the storage life of the fruit by at least a week. The overwrap and storage conditions, however, must not allow condensation inside the package, which would promote various decays. Appropriate perforation of the overwrap can prevent condensation, and a perforated overwrap still delays the onset of shrivel.

Ripe fruit tolerates low temperatures better than green fruit. Additionally, it may be possible to condition peppers to increase their tolerance of cold. Experimentally, tomatoes that have been treated with hot water tolerate chilling better than untreated ones, and the heat provides some control of postharvest pathogens. A portion of the improved export performance of pepper fruit after brushing with hot water may be due to increased resistance to chilling injury.

Selected References

Brecht, J. K., Chen, W., Sargent, S. A., Cordasco, and Bartz, J. A. 1999. Exposure of green tomatoes to hot water affects ripening and reduces decay and chilling injury. Proc. Fla. State Hortic. Soc. 112: 138–143.

Cantwell, M. 2000. Bell pepper. Produce Facts: Recommendations for Maintaining Postharvest Quality. On-line publication. Postharvest Technology Research and Information Center, Department of Pomology, University of California, Davis.

Fallik, E., Grinberg, S., Alkalai, S., Yekutieli, O., Wiseblum, A., Regev, R., Beres, H., and Bar-Lev, E. 1999. A unique rapid hotwater treatment to improve storage quality of sweet pepper. Postharvest Biol. Technol. 15:25–32.

McColloch, L. P. 1966. Botrytis rot of bell peppers. U.S. Dep. Agric. Mark. Res. Rep. 754.

(Prepared by J. A. Bartz)

Disease Caused by Angiosperms

Dodder

Dodder is a parasitic angiosperm that attacks a wide range of plants throughout North America and Europe. One of its common names, strangleweed, aptly describes its effect on host plants, including pepper. It is relatively uncommon on pepper but can severely stunt growth and depress pepper yields if it becomes established. Dodder is also a potential vector of pepper viruses when it produces shoots that grow from infected peppers to healthy ones.

Symptoms

Yellow or light green yellow strands of dodder entwine about pepper plants, growing out from initial infestation points until large areas of the host plants are covered (Plate 78). In-

fected pepper plants grow poorly, and their yield is reduced. As the host declines, dodder may produce white, pink, or yellow flowers, which can set seed.

Causal Organism

Several species of *Cuscuta* are commonly known as dodder. These plants produce thin but tough shoots that readily curl around a host. Their stems bear only small scales in place of true leaves, and the plants lack chlorophyll. They can produce abundant seeds only a few weeks after flowering.

Disease Cycle and Epidemiology

Dodder overseasons as seed in infested fields or mixed with pepper seed. Once the cropping season begins, dodder seeds germinate and produce shoots. If a shoot comes in contact with a pepper plant, the shoot produces haustoria, which penetrate stems and leaves of the host to reach vascular tissues, which serve as a source of water and nutrients for the parasite. The seeds produce no roots, and dodder dies if no suitable host is found within a few days of germination.

Once established, dodder vines grow and expand, engulfing pepper plants. Adjacent peppers become infected, and colonized plants may eventually die. After dodder seed sets, it may fall to the ground, where it can remain dormant until the next cropping season. Dodder seed can also contaminate commercial pepper seed lots.

Control

The best way to control dodder is to prevent its introduction into pepper fields. Pepper seed should be free of contaminating dodder seed. Vigilance must be maintained in the transplant production range, so that no dodder-infected material is moved to production fields. Areas of production fields with early infestations can be sprayed with a herbicide. The treatment results in the sacrifice of some pepper plants, but it may prove valuable in preventing further spread of dodder.

Selected References

Atsatt, P. R. 1973. Parasitic flowering plants: How did they evolve? Am. Nat. 107:502–510.

Kuijt, J. 1977. Haustoria of phanerogamic parasites. Annu. Rev. Phytopathol. 15:91–118.

Thodoy, M. G. 1991. On the histological relations between *Cuscuta* and its host. Ann. Bot. 25:655–682.

(Prepared by K. Pernezny)

Diseases Caused by Nematodes

Nematodes, sometimes called roundworms or eelworms, are members of the animal phylum Nematoda. Most are microscopic in size and have a characteristic vermiform, or worm-like, shape. They are abundant in soil and in freshwater and marine habitats, and many species are parasites of vertebrate animals, insects and other invertebrates, or plants.

Agricultural soils generally contain a complex community of many different kinds of nematodes, most of which feed on bacteria or fungi and are important in the decomposition and recycling of nutrients. Most soils contain low numbers of nematodes that feed as predators or omnivores. However, a substantial portion of the soil nematode community feeds directly on plant roots, sometimes causing damage or disease. Plant-parasitic nematodes in general are classified as endoparasites, which invade root tissue and spend a portion of their life cycle within plant tissue, or ectoparasites, which generally remain outside the roots. Plant-parasitic nematodes feed by means of a stylet (needle-like mouthparts), which they insert into root cells to remove cell contents.

The amount of damage caused by plant-parasitic nematodes depends on a number of factors. Probably the most important is the kind of nematodes present. Most soil samples contain plant-parasitic nematodes belonging to several different genera and species. Many of them, including spiral nematodes (*Helicotylenchus* spp.), ring nematodes (*Criconemella, Criconemoides,* and *Mesocriconema* spp.), and stunt nematodes (*Tylenchorhynchus* and *Quinisulcius* spp.), have not been shown to damage pepper and pose little threat to most agricultural crops. When a destructive nematode, such as *Meloidogyne incognita,* is present, the severity of the damage varies with the size of the nematode population. The selection of a cultivar for planting is also critical. Pepper cultivars range from highly susceptible to nematode-resistant.

Environmental factors also influence nematode damage and the expression of symptoms. For instance, nematodes feeding on or in roots can impede the uptake of water and nutrients by plants, resulting in wilting and other symptoms of water stress or nutrient deficiency. Water stress symptoms may be alleviated to some extent by rainfall or irrigation and may become worse with drought. Temperature, soil type, fertility, previous crops, the presence of other plant pathogens, and various other physical and biological factors and agricultural practices can affect symptom expression and the severity of nematode damage.

(Prepared by R. McSorley and S. H. Thomas)

Root-Knot Nematodes

Root-knot nematodes (*Meloidogyne* spp.) are by far the most serious nematode parasites of pepper. They occur worldwide, wherever pepper is grown, and have wide host ranges that include most solanaceous crops. Most reported root-knot damage has been due to *M. incognita* (Kofoid & White) Chitwood, which is present worldwide in warm climates, including tropical and subtropical regions and warm temperate locations, such as the southern United States and southern Europe. In cool temperate climates, *M. hapla* Chitwood sometimes causes damage during the summer.

Symptoms

Aboveground symptoms of root-knot nematode damage are typical of plants with damaged roots: stunting, wilting, chlorosis, nutritional deficiency, reduced fruit and leaf size, and low yield (Plate 79). Root-knot nematodes are endoparasites, and they induce the formation of giant cells in the root system, which are evident as galls, or knots, on the roots, a characteristic symptom (Plate 80). If a gall is examined closely, a light brown egg mass can be seen on the root surface, and a female nematode can be dissected from the root tissue beneath. Gaps due to seedling death may occur in fields of direct-seeded peppers where root-knot nematode infestations are severe. Collapsed seedlings often lack galls and show symptoms resembling injury due to salt, wind, or insects, but dissection of the radicle will reveal the nematode (Plate 81). Roots that are severely damaged by root-knot nematodes may be invaded by fungi and bacteria that cause further damage and root rot.

Causal Organisms

M. incognita is the most commonly reported nematode pest of peppers, but *M. hapla, M. javanica* (Treub) Chitwood, and *M. arenaria* (Neal) Chitwood also damage peppers in various locations.

The life cycles of these nematodes follow similar patterns, in which they pass through four juvenile stages before becoming adults. The first-stage juvenile molts in the egg, and the second-stage juvenile hatches from the egg. The second-stage juvenile is vermiform and 0.3–0.5 mm long, depending on the species. Root-knot nematodes are infective only as second-stage juveniles, which move through the soil to the roots of host plants. A nematode enters a root just behind the growing root tip and migrates between the undifferentiated root cells until it establishes a feeding site in the cells of the vascular system. At this point, it becomes a sedentary (immobile) endoparasite. Secretions from the nematode stimulate the formation of several giant cells, which provide nutrients for the nematode as it remains immobile and becomes swollen and sausage-shaped. Over the next few days, the nematode increases rapidly in size, molting into the third and then the fourth juvenile stages and finally into the adult stage, as either a pear-shaped female or a vermiform male. Males migrate out of the root without additional feeding. A female typically produces several hundred eggs, which are deposited in a gelatinous matrix at the posterior end of the body. The egg mass is apparent at the root surface once the rapidly growing body of the female ruptures the surface root cells. Parthenogenesis is common, and males are rare in most root-knot nematode populations.

Under optimum growing conditions for pepper in the field, the nematode life cycle from egg to egg takes about three to four weeks. However, the life cycle is strongly dependent on temperature and lengthens as the soil temperature decreases.

Host Races and Resistant Cultivars

Historically, four distinct host races of *M. incognita* and two races of *M. arenaria* have been recognized. The pepper cultivar California Wonder is one of the differential hosts used to distinguish species and races of root-knot nematodes. In the original differential host test, California Wonder pepper was considered to be a host of all four races of *M. incognita,* but not *M. javanica.* However, isolates (populations) of *M. javanica* that attack this cultivar have been identified, especially in Africa. Isolates of the same host race have been found to vary in their behavior and pathogenicity. These findings, along with the possible occurrence of multiple species and races of root-knot nematodes in the same field, cause difficulty in predicting host ranges and damage.

Numerous pepper cultivars have resistance to one or more species of root-knot nematodes. Resistance is known in both chili and sweet pepper cultivars. Resistance in recently developed cultivars is attributed to a single dominant gene (the *N* gene) or to this gene accompanied by a recessive gene. Cultivars expressing the *N* gene are resistant to both races of *M. arenaria, M. javanica,* and all four races of *M. incognita,* but they are susceptible to *M. hapla.* The recessive gene increases the temperature stability of root-knot nematode resistance in pepper.

Control

When available and adapted to local growing conditions, resistant cultivars may be a convenient means of managing root-knot nematodes in pepper. However, because of the variability among isolates of root-knot nematodes, cultivars with presumed resistance should be tested against local isolates before they are planted on a large scale. Cultivars lacking the recessive gene for resistance may perform poorly at high soil temperatures (28°C or above).

Chemical nematicides are often used for intensive production of high-value crops. Fumigant nematicides are often more effective against root-knot nematodes and produce more consistent results than granular nematicides. Broad-spectrum fumigants may offer protection against fungal pathogens and weeds as well.

A number of cultural practices may be helpful in managing root-knot nematodes in pepper crops. Sanitation is important at all stages, particularly in the use of nematode-free media and materials in the propagation of transplants. Removal of galled roots is another important sanitation practice. Soil solarization may be helpful in reducing nematode populations in the upper layers of the soil profile prior to planting. Crop rotation with nonhosts can be useful in suppressing root-knot nematode populations prior to pepper production. Grass, sorghum (*Sorghum bicolor*), or certain root-knot-resistant tropical legumes may be used for this purpose. Care is needed in the selection of rotational crops, however, because the different races, isolates, and species of root-knot nematodes have wide and varied host ranges. Finding a resistant crop that is adaptable to a local rotation may be a difficult task. For example, some cultivars of grain sorghum are nonhosts of *M. incognita* race 1 in the southeastern and south central United States but good hosts of *M. incognita* race 3 in west Texas and the southwestern United States. Furthermore, the presence of weed hosts in rotation crops can hinder the efficacy of the rotation. Yellow and purple nut sedges (*Cyperus* spp.), for example, allow root-knot nematode populations to increase in peppers and also protect the nematodes from chemical nematicides.

Selected References

McSorley, R. 1998. Alternative practices for managing plant-parasitic nematodes. Am. J. Altern. Agric. 13:98–104.

Netscher, C., and Sikora, R. A. 1990. Nematode parasites of vegetables. Pages 237–283 in: Plant Parasitic Nematodes in Subtropical and Tropical Agricultural. M. Luc, R. A. Sikora, and J. Bridge, eds. CAB International, Wallingford, U.K.

Potter, J. W., and Olthof, T. H. A. 1993. Nematode pests of vegetable crops. Pages 171–207 in: Plant Parasitic Nematodes in Temperate Agriculture. K. Evans, D. L. Trudgill, and J. M. Webster, eds. CAB International, Wallingford, U.K.

Taylor, A. L., and Sasser, J. N. 1978. Biology, Identification and Control of Root-Knot Nematodes (*Meloidogyne* Species). North Carolina State University Graphics, Raleigh.

Thies, J. A., and Fery, R. L. 2000. Characterization of resistance conferred by the *N* gene to *Meloidogyne arenaria* races 1 and 2, *M. hapla,* and *M. javanica* in two sets of isogenic lines of *Capsicum annuum* L. J. Am. Soc. Hortic. Sci. 125:71–75.

Thomas, S. H. 1994. Influence of 1,3-dichloropropene, fenamiphos, and carbofuran on *Meloidogyne incognita* populations and yield of chile peppers. J. Nematol. (Suppl.) 26:683–689.

(Prepared by R. McSorley and S. H. Thomas)

Sting Nematode

The sting nematode occurs in the southeastern United States, especially in soils with high sand content (more than 80% sand). It can cause severe damage to pepper and other vegetable crops.

Symptoms

Aboveground symptoms of sting nematode damage are similar to those caused by root-knot nematodes: stunting, chlorosis, wilting, and malnutrition. However, the sting nematode can be especially damaging to seedlings and transplants, occasionally killing young plants. In severely infested fields, plant growth may be uneven, plants may be extremely stunted, and gaps in the field may occur where seedlings have been killed. Symptoms are unevenly distributed in the field, since nematode distribution is uneven. The sting nematode is a large nematode that can severely damage or kill fine feeder roots. A lack of fine feeder roots or the presence of very short, stubby feeder roots, usually accompanied by necrotic lesions, is an important belowground symptom of sting nematode damage.

Causal Organism

Belonolaimus longicaudatus Rau is the common sting nematode in the southeastern United States (Fig. 6). Older literature may refer to it as *B. gracilis* Steiner. It is an amphimictic species, with both males and females common. It is vermiform in all stages of development. *B. longicaudatus* completes a gen-

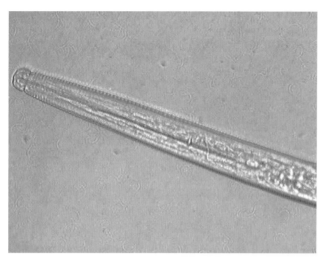

Fig. 6. Sting nematode.

eration in about one month during the growing season. At 2.0–3.0 mm in length, the adult sting nematode is much larger than most other nematodes in the soil. The sting nematode is generally confined to sandy soils, and it is believed that fine-textured soils with small pore size can restrict the movement of this large nematode. The sting nematode is an ectoparasite, but its unusually long stylet can penetrate deep into root tissue as it feeds. This feeding habit, along with its unusually large size, makes this nematode particularly damaging, even in small numbers. Isolates of *B. longicaudatus* from the southeastern United States vary in host range and pathogenicity.

Control

Many of the practices useful against root-knot nematodes, such as sanitation, solarization, and treatment with chemical nematicides, are also effective against the sting nematode. Granular nematicides may perform better against the sting nematode than against root-knot nematodes, since sting nematode populations do not build up as quickly. However, plant resistance to nematode ectoparasites is less common than resistance to endoparasites, and no pepper cultivars are known to have resistance to *B. longicaudatus*.

Because the sting nematode has a very wide host range, management by crop rotation can be difficult. Populations of *B. longicaudatus* increase rapidly on most sorghums and grasses, so these should be avoided where the nematode is present. Certain tropical legumes, such as hairy indigo (*Indigofera hirsuta*) and American joint-vetch (*Aeschynomene americana*), have been the most effective cover crops in rotation systems for managing the sting nematode.

Selected References

Christie, J. R. 1959. Plant Nematodes: Their Bionomics and Control. University of Florida, Gainesville.

Johnson, A. W. 1998. Vegetable crops. Pages 595–635 in: Plant and Nematode Interactions. K. R. Barker, G. A. Pederson, and G. L. Windham, eds. American Society of Agronomy, Crop Science Society of America, and Soil Science Society of America, Madison, Wis.

Rhoades, H. L., and Forbes, R. B. 1986. Effects of fallow, cover crops, organic mulches, and fenamiphos on nematode populations, soil nutrients, and subsequent crop growth. Nematropica 16:141–151.

Williams, K. J. O. 1974. *Belonolaimus longicaudatus.* Descriptions of Plant-Parasitic Nematodes, set 3, no. 40. Commonwealth Institute of Parasitology, St. Albans, U.K.

(Prepared by R. McSorley and S. H. Thomas)

Other Nematodes

Most nematode damage to pepper can be attributed to root-knot and sting nematodes, but other nematodes occasionally cause problems.

Stubby-root nematodes (*Paratrichodorus* and *Trichodorus* spp.) are worldwide in distribution and often occur along with the sting nematode in sandy soils in the southeastern United States.

The lesion nematode *Pratylenchus penetrans* (Cobb) Filipjev & Schuurmans Stekhoven is an important parasite in cool temperate regions.

The false root-knot nematode *Nacobbus aberrans* (Thorne) Thorne & Allen has been reported in chile pepper in Latin America.

A few reports have associated the reniform nematode, *Rotylenchulus reniformis* Linford & Oliveira, with pepper, but most sources consider pepper a nonhost of this nematode.

Other plant-parasitic nematodes associated with pepper include *Criconemella* spp., *Dolichodorus heterocephalus* Cobb, *Helicotylenchus dihystera* (Cobb) Sher, *Hemicycliophora arenaria* Raski, *Pratylenchus* spp., and *Radopholus similis* (Cobb) Thorne.

Symptoms

Aboveground symptoms of damage due to stubby-root, lesion, and false root-knot nematodes are similar to those caused by root-knot nematodes.

Stubby-root nematodes occasionally cause uneven growth and plant stands. Belowground, they damage fine feeder roots, leaving in short, stubby roots similar to those affected by the sting nematode.

Pratylenchus penetrans causes reduced root growth and small, dark brown lesions on roots. *Verticillium* has been associated with *P. penetrans* in pepper; its effects are additive, not synergistic.

N. aberrans induces the formation of small, beadlike galls on roots, similar in appearance to those caused by root-knot nematodes.

Causal Organisms

The stubby-root nematode associated with pepper in the southeastern United States is *Paratrichodorus minor* (Colbran) Siddiqi. *P. christiei* (Allen) Siddiqi and *Trichodorus christiei* Allen are considered synonyms of *P. minor* by some authorities. Stubby-root nematodes are ectoparasites that migrate upward through the soil to feed on root tips. *P. minor* is reported to be a vector of *Pepper ringspot virus* in South America and *Tobacco rattle virus* in peppers in Italy and the United States.

The lesion nematode *Pratylenchus penetrans* is an endoparasite that migrates into and within roots, causing characteristic lesions. Other members of the genus *Pratylenchus* have similar habits and may also be associated with pepper.

Lesion and stubby-root nematodes are vermiform in all stages of development. In contrast, mature females of *N. aberrans* are sedentary endoparasites, whose bodies increase in diameter as eggs are produced. Feeding by *N. aberrans* stimulates gall development, a characteristic symptom. The galls are smaller than those caused by root-knot nematodes. The female *N. aberrans* is also smaller than female root-knot nematodes.

Control

Control measures are similar to those for other plant-parasitic nematodes, with some important exceptions. *Paratrichodorus minor, Pratylenchus penetrans,* and *N. aberrans* all have wide host ranges, so that management by crop rotation is difficult. *P. minor* builds up on sorghum and possibly other grasses, but the host ranges for *P. minor* and *N. aberrans,* although fairly extensive, are not well known. No pepper cultivars are known to be resistant to any of these three species. *P. minor* can quickly recolonize soils that have been fumigated or solarized, presenting an additional difficulty in control.

Selected References

Hooper, D. J. 1977. *Paratrichodorus* (*Nanidorus*) *minor*. Descriptions of Plant-Parasitic Nematodes, set 7, no. 103. Commonwealth Institute of Helminthology, St. Albans, U.K.

Johnson, A. W. 1998. Vegetable crops. Pages 595–635 in: Plant and Nematode Interactions. K. R. Barker, G. A. Pederson, and G. L. Windham, eds. American Society of Agronomy, Crop Science Society of America, and Soil Science Society of America, Madison, Wis.

McSorley, R., Ozores-Hampton, M., Stansly, P. A., and Conner, J. M. 1999. Nematode management, soil fertility, and yield in organic vegetable production. Nematropica 29:205–213.

Netscher, C., and Sikora, R. A. 1990. Nematode parasites of vegetables. Pages 237–283 in: Plant Parasitic Nematodes in Subtropical and Tropical Agriculture. M. Luc, R. A. Sikora, and J. Bridge, eds. CAB International, Wallingford, U.K.

Potter, J. W., and Olthof, T. H. A. 1993. Nematode pests of vegetable crops. Pages 171–207 in: Plant Parasitic Nematodes in Temperate Agriculture. K. Evans, D. L. Trudgill, and J. M. Webster, eds. CAB International, Wallingford, U.K.

Shafiee, M. F., and Jenkins, W. R. 1963. Host-parasite relationships of *Capsicum frutescens* and *Pratylenchus penetrans, Meloidogyne incognita acrita,* and *M. hapla*. Phytopathology 53:325–328.

(Prepared by R. McSorley and S. H. Thomas)

Part II. Damage Caused by Arthropods

Arthropod damage to peppers can usually be attributed to members of a particular order on the basis of the type and location of the injury. Insects with chewing mouthparts consume or remove telltale portions of plants. Insects with rasping or piercing-sucking mouthparts sometimes cause symptoms similar to those of nutritional disorders or diseases. Feeding damage by some mite species can be mistaken for herbicide injury or viral diseases, because the tissue damage is disproportionate to that caused by the rasping mouthparts of the mites. Some insect pests of peppers also indirectly damage plants by transmitting viruses. Representatives of arthropod species whose direct feeding damage can be confused with biotic or abiotic disorders are briefly described below.

Aphids

Aphid pests of pepper include the green peach aphid, *Myzus persicae* (Sulzer); the melon aphid, or cotton aphid, *Aphis gossypii* Glover; the potato aphid, *Macrosiphum euphorbiae* (Thomas); and the glasshouse potato aphid, or foxglove aphid, *Aulacorthum solani* (Kaltenbach). Adults and nymphs are small (1.0–3.4 mm long), mobile, soft-bodied, and pear-shaped, ranging from light yellow to green to pink and black. Aphids have piercing-sucking mouthparts and feed on phloem in flowers and plant terminals and on the underside of leaves. Adults are winged or wingless. In warm climates females of many aphid species produce live nymphs directly, without an external egg stage. Nymphs look progressively more like adults with each instar and molt into the adult stage without passing through an immobile pupal stage. Their development is often completed within two weeks.

Symptoms

Most of the damage caused by aphid feeding in peppers is due to virus transmission, but direct feeding can also result in crop damage. Aphids excrete excess sugar water as honeydew through paired, tubular siphunculi, located toward the apex of the abdomen. This honeydew, like that of whiteflies, serves as a medium for the growth of sooty mold fungi, which reduce photosynthesis and detract from the appearance of fruit. Feeding on plant terminals and young leaves results in twisting and distortion of the leaves. Feeding by many aphids on somewhat older leaves results in leaf yellowing and curling. Large aphid colonies feeding in flower clusters can cause buds to be aborted.

Control

Cultural control of weeds and crop residues can reduce aphid populations. Rain and hot weather are also associated with reduced populations. Natural enemies eliminate many aphid colonies before they reach a damage threshold. Look for aphids on plant terminals, under young leaves, and around flowers in the upper half of pepper plants. Certain pesticides are effective in controlling particular species of aphids.

(Prepared by G. Nuessly)

Broad Mite

The broad mite, *Polyphagotarsonemus latus* (Banks), is only 0.15 mm long, invisible to most observers without magnification. Other common names refer to its body color or its hosts: it has been called the citrus silver mite, cotton mite, tropical mite, yellow mite, yellow tea mite, and white mite. In peppers, broad mites are likely to be found on terminals, flower buds, fruit, and young leaves.

Broad mites complete their development and begin producing eggs within four to five days. The clear, oblong, flattened eggs have six or seven rows of prominent white tubercles uniformly distributed across the surface. They are deposited in shallow depressions or along leaf veins, mostly under leaves or in other concealed locations. Six-legged, pear-shaped, white larvae, 0.1 mm long, emerge from eggs in about 30 hr and feed actively for 24 to 36 hr. After molting they enter a quiescent nymphal stage for another 24 to 36 hr before molting to the eight-legged adult stage. The nymphs are clear and pointed at both ends. The adults are initially translucent but become straw-colored or amber to green, with a central, dorsal white stripe.

Using its hook-shaped rear pair of legs and a special organ on the dorsum of the abdomen, the adult male carries a quiescent nymphal female on its back until the female matures. This behavior increases mating success and the lateral spread of infestation in pepper fields. Broad mites are also dispersed by phoresy—catching a ride on other arthropods, such as adult whiteflies. Females have a 30-hr preoviposition period, after which they produce an average of 1.5 to 2 eggs per day on peppers.

Symptoms

Although their size makes broad mites difficult to see, the damage they inflict is obvious and distinctive. Nymphs and adults feed on the undersurface of mostly younger leaves, flower buds, flowers, and fruit. Feeding damage in peppers is evident as bronzing, crinkling, cracking, and malformation of leaves and malformation and scarring of large areas of the fruit (Plates 82 and 83). Broad mite feeding also results in flower drop and the cessation of terminal growth and flower bud formation. Young pepper plants can be killed by prolonged feeding, and older plants recover slowly, taking three weeks to reestablish terminals and resume growth. Vegetative and reproductive output is never fully recovered after heavy damage has occurred.

Control

It is important to control broad mites before flower formation. They are difficult to eliminate once fruit has been formed, because they feed and deposit eggs under the calyx. Early sampling for broad mites should be concentrated at apical buds and the two youngest leaves. Infestations commonly begin near the borders of fields, particularly when they are located near other crop hosts, such as beans or citrus. Weeds that serve as reservoirs of broad mites, such as nightshades (*Solanum* spp.), should be controlled to reduce infestations. Whiteflies, which are destructive pests themselves, should be controlled in pepper and neighboring crops to reduce initial infestation with broad mites and their secondary spread.

Several insecticides and miticides provide effective control of broad mites. Two or three applications on a four- or five-day cycle may be necessary to control an infestation, because chemical treatments are not effective against broad mite eggs.

(Prepared by G. Nuessly)

Thrips

Some thrips species are pests throughout much of the world's pepper-growing regions, including the western flower thrips, *Frankliniella occidentalis* (Pergande); the melon thrips, *Thrips palmi* Karny; and the onion thrips, *T. tabaci* Lindeman. Other serious regional pests of pepper include *Chaetanaphothrips signipennis* (Bagnall), *Scirtothrips dorsalis* Hood, *T. hawaiiensis* (Morgan), and *T. parvispinus* Karny. Plant-feeding thrips are minute, slender-bodied insects, 0.5–3 mm long, and feed on pollen and the surface cells of terminals, flower buds, fruit, and leaves. Adult thrips have two pairs of narrow wings fringed with long hairs. They frequently hide and feed during the daylight hours within flowers, under the calyx of the fruit, under leaves along primary or secondary veins, and in other concealed locations. Thrips range from nearly white to various hues of yellow and orange, to light and dark brown, to black.

In warm climates, thrips develop from egg to adult in two to three weeks. Female thrips deposit eggs within the soft tissues of flowers, stems, young fruit, and leaves. Wingless larval thrips emerge to feed and move about actively through two instars. External wing buds are visible in the following two instars, the prepupa and pupa, in which the insect is inactive. This part of thrips development is usually passed in secluded locations in leaf litter or soil beneath pepper plants. Adults emerge from the pupal stage and return to plants to feed and reproduce.

Symptoms

Thrips infestations commonly begin near the borders of fields, particularly when they are located near other infested crop hosts. Pepper flower feeding is evidenced by scarring and sunken areas on the stigmatic surface, style, and anthers. If flower damage occurs before pollen is dehisced, fruit set is reduced, and the fruit is small and misshapen, with a low seed count. Heavy feeding on pollen can also lead to poor pollination and the production of distorted fruit. Damage to the surface of developing fruit results in curling and distortion of the calyx and suture-like scars. Later feeding on fully expanded fruit causes bronzing or silvering on the fruit shoulders and sides (Plate 84). Feeding on terminals and young leaves, particularly by *T. palmi,* results in the formation of small, distorted leaves (Plate 85), shortening of the internodes, and reduced flower production. Feeding on fully expanded leaves causes silvering and pitting along primary and secondary veins, usually on the lower leaf surface.

Some thrips also damage peppers by transmitting viruses, such as *Tomato spotted wilt virus.*

Control

Preplant measures to control thrips include planting resistant cultivars and mulching with plastic. It is important to control thrips pests early to reduce injury to the foliar and terminal region, which leads to decreased flower production and yield loss. In regions where foliage-feeding thrips are known to occur, search for thrips under leaves and in terminals, and look for injury symptoms. Flowers should be sampled frequently to ensure early detection of flower- and foliage-feeding thrips.

Foliar and soil drench insecticides are available for thrips treatment. However, species identification is critical to successful thrips control. Some thrips species are predaceous, but some are nonpest phytophagous species and serve an important role by providing early-season food for natural enemies of late-arriving pest species of thrips. Therefore, thrips species should be clearly identified before chemical treatment is applied.

(Prepared by G. Nuessly)

True Bugs

The most common pests of pepper in the order Heteroptera are stink bugs (family Pentatomidae) and plant bugs (family Miridae). Stink bug pests of peppers include the southern green stink bug, *Nezara viridula* (L.); the green stink bug, *Acrosternum hilare* (Say); the brown stink bug, *Euschistus servus* (Say); and others in the genus *Euschistus.* Plant bugs causing injuries that can be confused with disease symptoms include the tarnished plant bug, *Lygus lineolaris* (Palisot de Beauvois), and fleahoppers, *Halticus* spp.

All true bugs suck fluids from plant tissue through narrow mouthparts, which are commonly folded under the body at rest. They usually inhabit leaves, stems, flower buds, and fruit the in the upper regions of plants. These insects undergo incomplete metamorphosis in which the nymphs gradually look more like the adults with advancing instars. True bugs do not pass through an inactive pupal stage before molting into the adult stage. Both stink bug and plant bug adults are winged, with the basal part of the wing thickened and occluded and the distal part usually membranous and at least translucent.

Stink bugs are wide-bodied and shield-shaped, ranging from 1 to 2 cm long. Adults deposit groups of barrel-shaped eggs on leaves. Nymphs emerge from the eggs and frequently feed together in groups through at least the second instar. They gradually disperse as they increase in size. Stink bugs develop from eggs to adults in three to four weeks in warm climates. Nymphs of some species are brightly patterned with spots and contrasting colors, in contrast to the adults, which are usually mottled or a solid color.

Plant bugs attacking peppers are smaller (3–8 mm long) and narrower than stink bugs and are not shield-shaped. They deposit eggs singly in leaf or stem tissue. Nymphs do not usually live together in groups during early instars. Plant bugs reach maturity in 10 days to three weeks. Adults and nymphs are of similar coloring, usually mottled with various bands of beige to yellow or light to dark green over a dull yellow to green.

Symptoms

Stink bugs feeding on developing pepper fruit cause small, irregularly shaped, light-colored spots (Plate 86). Internally, the damage is characterized by light-colored granulated or corky areas in the fruit wall.

Flower buds damaged by tarnished plant bugs become discolored and then abort, and fruit set is reduced as a result. Tarnished plant bugs feeding directly on fruit cause light-colored spots similar to those due to stink bug feeding.

Fleahoppers feeding on leaves usually cause small, light-colored spots, but heavier feeding pressure may lead to wilting and leaf death. Large populations of fleahoppers can slow plant growth in the seedling stage.

Control

Preplant control of broadleaf weeds can help reduce developing populations of true bugs in and around pepper fields. Because there is a delay in the appearance of damage symptoms following fruit feeding, a pest population may no longer be at an unacceptable level when symptoms appear. Therefore, a decision to treat for true bugs should be based on insect density. Peppers in the seedling through flower bud stage should be sampled for fleahoppers in areas with a history of damage by this insect. Surveys for tarnished plant bugs and stink bugs should begin in the flower bud stage. Sampling can be facilitated the use of a sweep net or beat sheet to catch falling insects while the plants are lightly tapped or shaken. Insecticides can be used to control true bugs.

(Prepared by G. Nuessly)

Whiteflies

Whitefly pests of pepper include the greenhouse whitefly, *Trialeurodes vaporariorum* (Westwood), and the silverleaf whitefly, *Bemisia argentifolia* (Bellows & Perring). Whiteflies transmit pepper viruses, and they also inflict direct damage by feeding on pepper leaves.

Adult whiteflies are minute, soft-bodied insects, 1.2 mm long, with two pairs of waxy, white wings folded rooflike over the yellowish abdomen when they are at rest. Adults and nymphs have narrow piercing-sucking mouthparts and feed on sap from phloem. Eggs are deposited singly on the underside of leaves but may be arranged in groups or circular patterns. Crawlers emerge from eggs and settle a short distance away to feed. They do not move from their feeding sites through successive instars. Nymphs are 0.3–1 mm long, translucent to white, elliptical, flat, and scalelike and have several short hairs, the number and arrangement of which varies in different species. The red-eye or pupal stage is marked by the red eyes of the developing adult whitefly within the exoskeleton of the last nymphal instar. Adults emerge to freely move about, reproduce, and feed. They develop within four weeks and may move into pepper fields by the thousands from nearby infested fields.

Symptoms

In processing sap to extract nutrients, whiteflies produce copious amounts of honeydew. This sticky residue provides food for sooty mold fungi, which reduce photosynthesis and detract from the appearance of the fruit. A large population of whiteflies on pepper can cause plant stunting, reduced growth rate, defoliation, and reduced yield.

Control

Cultural practices, such as crop destruction immediately following harvest, control of weed hosts and volunteer crop plants, maintaining host-free periods, and planting at least 1 km upwind along the predominant wind direction, help considerably to control whiteflies. Predation and parasitism provide some level of control, particularly in greenhouses with augmentative releases of natural enemies of whiteflies and strict control of pest immigration. Field margins should be scouted to detect early-season infestations. Yellow sticky cards can be used to indirectly sample populations of whiteflies and their natural enemies. Pesticides can provide good control, but precautions should be taken to avoid treatment schedules that select for resistance by exposing a whitefly population to repeated applications of insecticides of the same class.

(Prepared by G. Nuessly)

Part III. Abiotic and Physiological Disorders

Abnormal Fruit Shape

Symptoms

The fruit of any bell pepper cultivar will vary from blocky to oblong, but the amount of variation in shape is greater in some cultivars. In extreme cases, the fruit is severely flattened and much broader than long (Plate 87); such a fruit is colloquially known as a cookie or a cheese pepper.

Causes

Temperature is the primary environmental factor causing variations in the shape of pepper fruit. If plants are exposed to high temperatures (32°C during the day and 21°C at night) for several days or more during the preflowering period, when flower buds are just forming, the fruit that develops from those flowers tends to be more blocky than normal and tends to have more segments or locules. Under cool conditions (day and night temperatures of 15 and 5°C) during flowering, many flowers are lost. Some flowers form fruit and set a few seeds, but the fruit will be nearly seedless and abnormally flattened.

Control

Growers can try to choose a transplanting date when extreme heat is not expected during the first month after transplanting. Little can be done to avoid fruit malformation induced by cool weather, but some varieties are less severely affected by the disorder.

(Prepared by H. C. Wien and T. A. Zitter)

Blossom-End Rot

Blossom-end rot is a noninfectious, physiological disorder caused by calcium deficiency in the blossom end of developing pepper fruit. The disorder is aggravated by any factor that reduces the uptake of calcium by the plant. One of the most important factors in the development of blossom-end rot is the availability of water. Since calcium is a relatively immobile element, fluctuations in water availability, even for short periods, can result in deficiency symptoms. Blossom-end rot is particularly severe when heavy rains early in the growing season are followed by drought or a lengthy dry period prior to irrigation. As the weather warms and plants begin to grow rapidly, their requirements for water and calcium increase. Fast-growing plants are highly susceptible to blossom-end rot, and therefore the disorder is most common among the fruit set earliest. Other factors contributing to blossom-end rot include high soil salinity, root pruning, and heavy application of nitrogen or magnesium.

Symptoms

Blossom-end rot first appears as a small, water-soaked, light brown spot on the distal end of a developing fruit. On chili peppers, and to a lesser extent on bell peppers, the spot may appear off to the side of the blossom end (Plate 88). As the fruit grows, the spot enlarges, until eventually it covers as much as half the fruit. Over time, the lesion becomes sunken and leathery (Plate 89) and ultimately may appear straw-colored and papery. Fungi and bacteria may invade the weakened tissue, causing it to turn black or appear watery. Affected fruit ripens faster than unaffected fruit.

Control

This disorder is best controlled with adequate water and fertilizer management. Sufficient but not excessive amounts of water and fertilizer should be provided through the growing season. A preplant soil test is recommended to determine base levels of calcium in the soil.

Soil amendments with lime and other calcium fertilizations have been effective in reducing the incidence of blossom-end rot in some locations. Likewise, foliar sprays with anhydrous calcium chloride during the growing season may help to alleviate the disorder, but the results of this treatment have been erratic.

(Prepared by N. Goldberg)

Color Spotting

Symptoms

Color spotting is a fruit disorder in which round, dark spots, 2 to 6 mm ($^1/_{16}$ to $^1/_4$ inch) in diameter and slightly sunken below the surface, form on fruit (Plate 90). The symptoms show most clearly in yellow and red fruits and are hardest to see at the mature green stage of fruit growth. Spots of similar size, but yellow, were described on greenhouse-grown bell peppers in Israel, appearing first in mature green fruit turning color. Pale to dark green pits were reported on maturing bell pepper fruit in Australia.

Causes

Spots of different colors may have different causes, and therefore this disorder occurs sporadically.

Black spotting appears to be caused by exposure of fruit to temperatures just above freezing for several nights. It appears to be a form of chilling injury affecting fruit in the field when the weather is cool around the time of harvest and under storage conditions after harvest.

The yellow spotting of greenhouse-grown peppers in Israel was evident when fruiting plants were exposed to temperatures of 30–35°C, with reduced light and high nitrogen fertility.

The green pitting described in Australia occurred at normal growing temperatures. In both the Israeli and Australian cases, pitting was associated with higher than normal calcium content of the fruit, especially in the pitted area.

Marked differences in the incidence of color spotting have been found in different cultivars.

Control

Selection of cultivars less susceptible to color spotting may be the best way to avoid the disorder. Avoiding conditions that induce spotting—such as chilling, in the case of black spotting, and high temperatures, in the case of yellow spotting—will also help. Maintaining an adequate but not excessive level of calcium in the soil may help prevent color spotting.

(Prepared by H. C. Wien and T. A. Zitter)

Flower and Flower Bud Drop

Symptoms

An abiotic disorder can cause flower buds, open flowers, and sometimes young pods to fall off growing pepper plants (Plate 91). In extreme cases, a plant may lose all its reproductive structures, including tiny flower buds, so that flowering cannot resume for several weeks or more.

Causes

The principal cause of flower and flower bud drop is high temperature (day and night temperatures of 32 and 21°C or higher) sustained for five days or more. Less often, damage from sucking insects, such as the tarnished plant bug and particularly the pepper weevil, can lead to bud drop. Loss of open flowers can also occur when flowering plants are exposed to cool weather (day and night temperatures of 15 and 5°C), during which the growth of pollen tubes is slowed, so that normal pollination is prevented. The disorder can also be caused by various stresses, including insufficient water, strong wind, high relative humidity, nutrient deficiency or toxicity, and nutrient imbalances.

Control

Avoiding high temperatures by adjusting the planting date or moving to a cooler location may be feasible for some growers. The best management practice is to ensure adequate water and fertilization during flowering. Some varieties of bell pepper are less susceptible to flower bud drop, so variety trials conducted during problem periods may be useful. Good insect control may be helpful if sucking insects are suspected to be the cause. Flower loss due to cool weather rarely lasts long enough to cause sustained reduction of flowering, so the best way to control this disorder is to wait for warmer conditions.

(Prepared by N. Goldberg,
H. C. Wien, and T. A. Zitter)

Fruit Cracking

Symptoms

Fruit cracking is an abiotic disorder in which cracks form in the skin of the fruit, varying in length from 3 mm to several centimeters ($^1/_8$ inch to several inches) and in depth, from barely breaking the surface to extending into the fruit wall. These cracks appear in fruit that is mature green or older. In the jalapeño type of pepper, shallow surface cracks are a type characteristic and are considered normal.

Causes

Fruit cracking is infrequent in bell peppers, especially when they are grown outdoors. It is more common in greenhouse culture, if nighttime relative humidity is near saturation. Under those conditions, the fruit imports so much fluid that its skin is not able to accommodate the expansion, and the skin cracks. Exposure of large fruit to the sun aggravates the cracking. Removal of leaves increases exposure to the sun and also increases the amount liquid taken up by the fruit.

Control

In greenhouse production, avoid high humidity at night. Maintain good foliage cover over the fruit, and select cultivars that are less susceptible to fruit cracking.

(Prepared by H. C. Wien and T. A. Zitter)

Hail Injury

Severe hail can leave foliage tattered (Plate 92) and fruit pockmarked (Plate 93). Damaged tissue may become chlorotic or necrotic in response to the injury. Wounds created by hail are often invaded by fungi and bacteria, which subsequently cause fruit decay.

(Prepared by N. Goldberg)

Salt Injury

Excessive concentrations of salts in soil or irrigation water can cause substantial crop losses. Peppers are relatively sensitive to salts; yield loss begins when electrical conductivity (EC) is over 1.5 dS/m. In New Mexico, 50% yield loss occurs at EC 5.8 dS/m, with an additional 12.6% reduction in yield for every additional unit increase in EC. Plants of all ages are susceptible to salt injury, but the damage is generally more severe in seedlings than in older plants. Seedlings can be severely stunted or killed (Plate 94). In direct-seeded peppers, salt injury together with seedling diseases or wind damage can result in long skips in the plant stand (Plate 95), and stand losses of over 50% have been observed. Symptoms of salt injury in mature plants are burned root tips, marginal leaf necrosis, wilt, and defoliation (Plate 96).

The injury often develops after light rain or light irrigation washes salts into the root zone. The problem cannot be eliminated, but the damage can be minimized by cultural practices that move salts away from the plants and the root zone. For example, furrow-irrigated plants should be planted to one side of the bed, because this irrigation method tends to concentrate salts in the center of the bed. Conversely, drip-irrigated plants should be planted in the center of the bed, because this irrigation method tends to deposit salts at the sides of the row. Regardless of the irrigation method chosen, enough water should be applied to help leach salts into the soil. Seedling damage may be reduced by planting in preirrigated beds and capping the row. When the cap is removed, salt is dispersed in the furrow or on the sides of the beds.

Pepper cultivars have varying degrees of salt tolerance.

(Prepared by N. Goldberg)

Sunscald

Symptoms

Sunscald occurs on foliage exposed to intense sunlight at high temperatures (Plate 97). It also causes sunken, dead areas to form on fruit, on the side exposed to the sun. Affected areas are light-colored, soft, and wrinkled. The damaged tissue eventually turns whitish tan and papery in texture (Plate 98). Affected areas are frequently white but may be discolored if they become infected by fungi (Plate 99). Fungi and bacteria that invade damaged tissue contribute to the decay of the fruit. Fruit symptoms are easily confused with those of blossom-end rot, except that sunscald will only occur on the side of the fruit exposed to the sun, whereas blossom-end rot lesions may form on unexposed areas.

Causes

Sunscald occurs on pepper fruit that is directly exposed to intense sunlight. It often occurs when shaded fruit is suddenly exposed to the sun. The exposed tissues become so hot that they become damaged. The symptoms are most common on plants with little foliage cover, such as plants inadequately fertilized with nitrogen. In addition, defoliation or prolonged wilting caused by other diseases, such as powdery mildew, bacterial spot, Phytophthora blight, Verticillium wilt, and root-knot nematodes, can contribute to an increase in the incidence of sunscald. Alternately, pepper plants may become top-heavy with fruit, and branches may break or fall over in a rainstorm, exposing fruit to the sun. Such breakage can also occur when branches are handled too roughly during harvest.

Control

To help develop an adequate foliage cover over the fruit, adequate levels of nutrients, particularly nitrogen, are essential. In addition, avoidance of drought stress through irrigation promotes good leaf production. Plants can be supported with stakes or horizontal wires or string running along the rows or at the edge of beds. Diseases that cause defoliation or wilting should be controlled.

(Prepared by N. Goldberg,
H. C. Wien, and T. A. Zitter)

Wind Injury

Strong wind damages pepper plants in several ways. Damage can result from rapid desiccation of the foliage or hypocotyl. When this happens to young seedlings, the plants fall over and die (Plate 100). Desiccated foliage on older plants may wilt beyond recovery. Wind injury may also result in physical damage to foliage (Plate 101) or broken stems or branches. Seedlings can be damaged when the wind whips them back and forth (Plate 102) and typically snap off where callus tissue forms at the soil line following injury. Wind damage can be intensified by blowing sand, which causes necrotic spots to form on foliage.

One method of protecting seedlings from strong winds is to plant in the stubble of small grains (Plate 103). Additionally, windbreaks planted around the edges of a field may provide protection for plants growing in windy areas. This is a common practice in bell pepper production in the southeastern United States.

(Prepared by N. Goldberg)

Selected References for Abiotic and Physiological Disorders

Hamilton, L. C., and Ogle, W. L. 1961. The influence of nutrition on blossom-end rot of pimiento peppers. Am. Soc. Hortic. Sci. 80:457–461.

Higgins, B. B. 1925. Blossom-end rot of pepper (*Capsicum annuum* L.). Phytopathology 15:223–229.

Ivanoff, S. S. 1944. Guttation-salt injury on leaves of cantaloupe, pepper, and onion. Phytopathology 34:436–437.

Leyendecker, P. J., Jr. 1950. Blossom-end rot of pepper (*Capsicum annuum* L.) in New Mexico. Phytopathology 40:746–748.

Monette, S., and Stewart, K. A. 1987. The effect of a windbreak and mulch on the growth and yield of pepper (*Capsicum annuum* L.). Can. J. Plant Sci. 67:315–320.

van der Beek, J. G., and Ltifi, A. 1991. Evidence for salt tolerance in pepper varieties (*Capsicum annuum* L.) in Tunisia. Euphytica 57:51–56.

Villa-Castorena, M., Ulery, A. L., and Catalan-Valencia, E. 2000. Salinity and nitrogen effects on nutrient uptake by chile pepper plants. Am. Soc. Agron. Meet., Agron. Abstr.

Part IV. Herbicide Injury

Numerous herbicides are registered for weed control in peppers (bell peppers and other types). They rarely cause significant economic injury, in which yield loss due to injury exceeds the yield increase due to reduced interference from weeds as a result of herbicide application. Some herbicides occasionally produce visible symptoms on pepper plants. Economic injury may occur if a herbicide is applied in a manner other than that described on the product label—for example, if it is applied at a rate higher than recommended or if the timing of applications is incorrect. Significant economic injury is most likely to occur when pepper plants are exposed to herbicides that are inappropriately or accidentally applied, herbicides that drift onto pepper plants from nearby fields where they were applied, or herbicides that were applied to a previous crop and persist in the soil long enough to injure peppers planted later in the same field.

Diagnosing injury caused by herbicides is difficult and often confusing. Symptoms caused by a specific herbicide can vary, depending on the dosage and the stage of plant growth at the time of exposure. Large, mature plants may exhibit few or no visible symptoms after exposure to a herbicide that could kill a smaller, less mature plant at the same dosage. Also, many symptoms caused by viruses, nutrient disorders, insect feeding, environmental conditions, and air or water pollution can be confused with symptoms of injury caused by certain herbicides. One should be cautious when observing symptoms of plants that have been removed from the field. Observations of the pattern of injury in the field and other species of plants affected are critical to the correct identification of herbicide injury.

Three steps should be taken to determine whether a herbicide is likely to have caused symptoms observed on peppers:

Step 1. The symptoms should be consistent with those known to be caused by a particular herbicide or members of a particular herbicide family, in the personal experience of the observer and as reported by authoritative sources. The observer should also inspect weed species in the field and determine whether they have symptoms similar to those of the pepper plants. If it can be determined that the symptoms are possibly attributable to a particular herbicide or members of a particular herbicide family then the observer should proceed to Step 2.

Step 2. The diagnostician should determine whether the suspected herbicide or a member of the suspected herbicide family was used in the field and whether it was used in a manner that could have exposed pepper plants to it. This information may be difficult to obtain or unavailable at the time of plant inspection. With time and effort, however, the source of herbicide exposure can usually be determined. Even if the source cannot be immediately identified, the observer should proceed to Step 3.

Step 3. Chemical analysis of symptomatic plants or soil may allow a positive identification of the herbicide that caused the injury. However, the symptoms on pepper plants must be consistent with those caused by a herbicide detected by chemical analysis. Plants with mild symptoms may not contain enough of a suspected herbicide to test positive; however, economic injury is less likely in this case. With the current capability of analytical methods for herbicide detection, plants that are severely injured or dead usually contain herbicide in amounts sufficient for detection. Many commercial laboratories and some field-testing kits can analyze samples to detect most commonly used herbicides. Several herbicides may cause similar symptoms; therefore, more than one chemical analysis may be needed. In some circumstances, the cost of testing may be prohibitive. However, without chemical analysis, the source of the injury cannot be positively identified.

Symptoms Caused by Herbicides Registered for Use on Peppers

Generally, few of the herbicides registered for use on peppers cause symptoms. Occasionally, symptoms may be visible when one these herbicides is used in a manner other than that described on the product label (Table 2). Symptoms may occur when a registered herbicide is applied at a dosage higher than recommended, with improper timing, or in combination with additives contraindicated on the label. Care must be taken during application to avoid excessive overlap of spray swaths and double applications at the edges of fields and the ends of rows. Some pepper cultivars may be more sensitive than others to registered herbicides. Before using a new herbicide, growers should consult with the manufacturer to determine whether cultivars vary in sensitivity to the product; similarly, before planting a new cultivar, they should consult with the seed dealer to determine its sensitivity to herbicides.

Symptoms Caused by Accidental Application of Phytotoxic Herbicides

Accidental application of phytotoxic herbicides to peppers most commonly occurs when spray equipment is contaminated with a product applied earlier or when the wrong product is added to a spray tank. Symptoms caused by sprayer contamination are usually most intense in the area of the field where the first tank of a solution or suspension was sprayed. The symptoms are less obvious or absent in areas sprayed later. In the rare event that a fertilizer or another pesticides is contaminated with a small amount of a phytotoxic herbicide, symptoms of herbicide injury will be uniformly distributed across the field. Fertilizers and other pesticides should be stored separately from herbicides. The most severe injuries are usually caused when a phytotoxic herbicide is mistakenly added to a spray tank instead of the desired pesticide. Again, symptoms will be uniformly distributed across the field.

Symptoms Caused by Drift of Phytotoxic Herbicides

Herbicides can move into pepper fields either as vapor drift or as spray drift from nearby fields.

Few herbicides are volatile enough to injure peppers as a result of vapor drift. Ester formulations of the phenoxy and picolinic acids (for example, 2,4-D and picloram, respectively) are the most common causes of vapor drift injury to peppers. Vapor drift occurs most readily at high temperatures, when the soil and plant surfaces are wet from spraying, and when the air

TABLE 2. Response of Pepper to Various Herbicides

Herbicide	Symptoms	Remarks

Acetyl coenzyme A carboxylase inhibitors
  Fluazifop-P
  Sethoxydim
  Clethodim
	Necrotic spots on leaves are occasionally caused by various adjuvants (usually crop oil) used with these herbicides.	These herbicides are registered for use on peppers and are considered safe. Symptoms rarely occur and are transitory.

Acetolactate synthase inhibitors
  Sulfonylureas (partial listing)
    Chlorsulfuron
    Metsulfuron
    Bensulfuron
    Nicosulfuron
    Halosulfuron (Plate 104)
  Imidazolinones (partial listing)
    Imazaquin
    Imazethapyr (Plate 105)
    Imazapyr
    Imazameth
  Others (partial listing)
    Pyrithiobac (Plate 106)
    Diclosulam
	Most obvious foliar symptoms are in terminal growth. Symptoms vary, depending on exposure and sensitivity, from blackening or yellowing of terminal growth to reduced leaf size and leaf distortion. Plants may be stunted as a result of reduced internode length. Root growth may be inhibited by soil exposure.	Sensitivity to these herbicides varies greatly. Some products may soon be registered for use on peppers. Symptoms can be caused by foliar exposure or soil exposure.

Photosynthesis inhibitors
  Triazines (partial listing)
    Atrazine (Plate 107)
    Simazine
  Triazinones
    Hexazinone (Plate 108)
    Metribuzin
  Substituted ureas (partial listing)
    Diuron
    Linuron
  Uracils
    Bromacil
    Terbacil
	Foliar contact may cause irregular chlorosis or necrosis. Symptoms from root uptake are veinal or interveinal chlorosis, becoming most obvious in older leaves first. Marginal chlorosis followed by necrosis may be observed in leaves exposed to these herbicides at high rates of application.	All of these herbicides remain active in the soil and may persist long enough to injure subsequently planted peppers. Foliar exposure can occur as a result of drift. At low levels of exposure symptoms may not be evident, but the growth of exposed plants will be reduced, compared to that of unexposed plants.

Tubulin inhibitors
  Dinitroanilines (partial listing)
    Trifluralin
    Pendimethalin
  Benzoic acid
    DCPA (chlorthal)
  Benzamide
    Isoxaben
	Emergence of seedlings is delayed or prevented. Root tips may be swollen, and root growth is reduced, resulting in overall plant stunting. Stem swelling or cracking may occur at the soil surface.
	
	Injury due to DCPA is not likely; symptoms have not been observed.
	
	Symptoms of injury due to isoxaben have not been observed.
		These herbicides have little or no foliar activity. Injury is not expected when they are used on peppers according to the instructions on the product label.

General cell division inhibitors
  Thiocarbamates (partial listing)
    EPTC
    Vernolate
    Molinate
    Butylate
  Amides (partial listing)
    Alachlor
    Metolachlor
    Diphenamid
    Napropamide
  Others (partial listing)
    Bensulide
	Seedlings may not emerge from the soil, and root growth may be reduced. General stunting may occur, with crinkling or slight malformation of the first leaves to form.	Bensulide and metolachlor rarely cause symptoms when used on peppers according to the instructions on the product label.

(continued on next page)

is calm, allowing a high concentration of vapor to accumulate. Using salt formulations of these herbicides will greatly reduce vapor drift. Generally, symptoms caused by vapor drift are most severe in low-lying areas where air containing the vapor may settle. Symptoms are less obvious as the distance from the target field increases.

Spray drift is the movement of spray droplets from a target area to a nontarget area, such as a pepper field. Wind speed, droplet size, and height of the spray boom during application determine the extent of spray drift. Commonly used herbicides that can cause symptoms on peppers as a result of foliar contact due to spray drift include 2,4-D, dicamba, picloram, quinclorac, glyphosate, glufosinate, paraquat, acetolactate synthase (ALS) inhibitors, and protox inhibitors (contact herbicides such as oxyfluorfen and carfentrazone). Symptoms are most severe on pepper plants closest to the target area and decrease in severity as the distance from the target area increases. Spray drift can be avoided by spraying when the wind is blowing away from pepper fields, leaving unsprayed buffer zones between the target area and pepper fields, applying sprays with a large droplet size, using drift control additives, and placing the spray boom as close as possible to the surface to be sprayed.

Symptoms Caused by Carryover of Residual Herbicides

Soil-active herbicides used for weed control in other crops prior to the establishment of peppers may persist long enough to cause symptoms in peppers when they are subsequently planted. Products that provide season-long weed control tend to persist in the soil, and restrictions on crop rotation described

TABLE 2. (*continued*)

Herbicide	Symptoms	Remarks
Carotenoid pigment inhibitors Norflurazon (Plate 109) Clomazone (Plate 110) Amitrole	These herbicides cause loss of chlorophyll in leaf tissue, resulting in a white (bleached) to yellowish appearance. Clomazone causes interveinal bleaching of leaves, whereas norflurazon causes veinal bleaching.	Clomazone rarely causes symptoms when used on peppers according to the instructions on the product label. Norflurazon and amitrole can cause injury to peppers planted in fields where they were previously applied.
Contact action herbicides Diphenyl ethers Acifluorfen Lactofen Oxyfluorfen (Plate 111) Fomesafen Triazinone Carfentrazone (Plate 112) Sulfentrazone Bipyridiliums Paraquat (Plate 113) Diquat Others MSMA Bromoxynil Bentazon (Plate 114) Propanil	These herbicides cause rapid injury to foliage. Symptoms vary from rapid wilting and desiccation to yellow spotting, with the centers of the spots becoming necrotic. Plants do not die unless they are sprayed directly at a high rate of application. Translocation in the plant is limited, and symptoms develop only on foliage in contact with the herbicide. Carfentrazone causes irregularly shaped necrotic spots on contacted leaves, but new growth will appear normal. Paraquat causes circular yellow spots surrounded by a halo. The spots become necrotic and may coalesce over time. If new growth is produced, it will be normal. MSMA may cause a more general yellowing. Bentazon causes irregular yellowing or whitening of leaves and, in cases of severe injury, leaf necrosis, but new growth will be normal.	Spray drift is the major cause of injury by all of these herbicides.
Glutamine synthetase inhibitor Glufosinate	Foliar symptoms develop more rapidly than those due to glyphosate but more slowly than those due to paraquat. Wilting and desiccation of foliage are visible after a few days. The youngest tissue exhibits the most severe symptoms.	Regrowth of plants affected by glufosinate will appear normal.
EPSP synthase inhibitor Glyphosate (Plates 115 and 116)	Foliar contact causes chlorosis of terminal leaves, but the symptoms develop slowly, appearing several days after exposure. Plants may die, and new growth may be malformed.	Glyphosate is inactive in mineral soils but can be absorbed by roots of plants in an organic growing medium such as bark or vermiculite. Malformation of leaves and fruit can occur as a result of this type of exposure.
Hormone growth regulators Phenoxy acids (partial listing) 2,4-D 2,4-DB MCPA Benzoic acid Dicamba (Plate 117) Pyridines Picloram Triclopyr Clopyralid Quinolinecarboxylic acid Quinclorac (Plate 118)	These herbicides may cause epinasty in peppers. Common symptoms are strapping, cupping, and feathering of leaves, petiole twisting, and stem bending. Symptoms are most evident on leaves that were forming at the time of exposure or formed later. Phenoxy acids can cause stem swelling and cracking and the formation of callus tissue. Symptoms vary, depending on the herbicide and the level of exposure.	Foliar contact from herbicide drift is the most common form of exposure, and symptoms can develop at exceedingly low levels of exposure. Ester formulations of the phenoxy acids and pyridines are more likely to volatilize than salt formulations. All of these herbicides are active in the soil and may be taken up by pepper roots, causing symptoms similar to those due to foliar exposure.

on the product labels should be followed. Occasionally, application at a rate higher than that described on the label, spray boom overlap, or cool and dry conditions may cause a herbicide to persist longer than normal and injure a subsequent pepper crop. Symptoms may vary with soil texture (they are generally less obvious as the clay content increases and more obvious as the sand content increases), compaction of soil in areas of a field, and low or high spots. Spray boom overlap during applications made the previous season may give rise to an obviously traceable pattern of symptoms. Factors such as the pH and organic matter content of the soil can affect the persistence of many herbicides. The herbicides most likely to persist and cause symptoms on peppers are the triazines (e.g., atrazine and simazine), substituted ureas (e.g., diuron), norflurazon, sulfonylureas (e.g., chlorsulfuron and metsulfuron) and imidazolinones (e.g., imazethapyr and imazapyr).

Table 2 and Plates 104–118 describe and illustrate symptoms caused by various herbicides on pepper plants. However, this presentation should not be the only source information for diagnosing herbicide injury. The diagnostician should also rely on corroborating evidence and consider other factors that cause similar symptoms.

(Prepared by J. Schroeder and P. A. Banks)

Part V. Nutritional Disorders

Pepper requires optimal amounts of essential elements in order to produce the greatest yield of high-quality fruit. Essential nutrients can be supplied by the soil, with its native fertility, or by fertilizers. Inadequate nutrition of pepper plants is manifested in reduced plant growth and fruit yield. Nutrient deficiencies can also lead to certain fruit disorders or impaired fruit quality. Some of these disorders are direct effects of a nutrient deficiency, while others may be indirect effects of environmental factors such as drought or low temperature, root diseases, or nematode damage to roots.

Nitrogen

Pepper plants under stress from low levels of nitrogen (N) exhibit reduced growth, producing leaves and fruit that are smaller than normal. N deficiency is manifested as a general yellowing of entire leaves and an overall diminution of the green color of the fruit, rendering it less acceptable for market. The fruit of N-deficient plants is thin-walled, and bell peppers produce less fruit with the desired blocky, four-lobed shape. The deficiency affects older leaves first, since the nutrient is translocated from the older leaves of plants under N stress, to sustain growth in the upper parts of the plant. The most recently matured leaves of N-deficient pepper plants at early fruit setting will have total N concentrations of less than 2.5% (dry weight basis). Deficiencies in commercial plantings typically result from inadequate N fertilization. In coarse sandy soils, N can be lost from the root zone as a result of leaching due to heavy rainfall or excessive irrigation. If diagnosed early enough, the deficiency can be corrected by timely application of N fertilizer as side-dressing or in irrigation water.

Potassium

Potassium (K) deficiency causes a leaf disorder sometimes referred to as bronzing. Older leaves of K-deficient plants turn tan and then brown at the margins. In cases of prolonged deficiency, the leaves become necrotic at the margins, with some interveinal yellowing or tanning, giving rise to the term *bronzing* (Plate 119). The most recently matured whole leaves of such plants will have K concentrations of less than 2.0% (dry weight basis). Plants under K stress are smaller than normal (Plate 120) and produce fewer and smaller fruit with thinner walls. K is mobile in the plant; hence the deficiency occurs initially on the lower leaves and advances to the middle leaves of the plant. Deficiencies are most often the result of inadequate fertilization. K is mobile in coarse, sandy soils, and deficiencies can result from leaching due to heavy rainfall or excessive irrigation. Like N deficiency, K deficiency can be corrected by timely application of K fertilizer as side-dressing or by fertigation.

Phosphorus

Phosphorus (P) deficiency is rarely observed in commercial pepper crops, because most agricultural soils contain adequate reserves of this nutrient, and most growers add it to fertilizers. P is mobile in the plant, so that a deficiency affects older leaves first. The deficiency is characterized by reduced growth of the plant and leaves that are smaller and darker green than normal. The most recently matured leaves exhibiting the deficiency will have P concentrations of less than 0.15%. P deficiency in the field is most likely to occur in a crop planted on newly cleared land that has not been in agricultural production and to which very little P has been applied. Provision for adequate P nutrition should be made before planting, since post-plant applications are typically not effective.

Calcium

Calcium (Ca) deficiency in pepper can lead to blossom-end rot, a fruit disorder in which a sunken, dark lesion develops at the distal end of the fruit. Blossom-end rot lesions range from a few millimeters to several centimeters in diameter, depending on the size of the fruit and the severity of the Ca deficiency. The disorder can develop even though there is ample Ca in the soil in which the crop is grown. In addition, there might be adequate concentrations of Ca in the leaves of affected plants. This apparent anomaly arises because Ca is translocated in the plant to organs from which transpiration is greatest, such as leaves, thus largely bypassing the fruits. Under mild drought stress or in soils with high osmotic potential, Ca movement will be preferentially directed toward the leaves. Young developing fruit is most vulnerable to Ca deficiency, because it is expanding rapidly. Ca is not highly mobile in phloem, and young fruit can quickly suffer tissue collapse from a deficiency of the nutrient. Leaf symptoms of Ca deficiency are rarely observed in the field. The application of lime and many fertilizer materials and fertilizer filler materials provide Ca to the crop. In many growing regions, the nutrient is supplied in irrigation water drawn from wells in limestone aquifers. Stunting and necrosis of the youngest leaves and growing points occurs at the top of the plant, in the newest growth. Damaged leaves sometimes take on a curled, straplike appearance. Vegetative deficiency symptoms are most likely to be exhibited in plants whose younger leaves have Ca concentrations of less than 0.5%. Blossom-end rot can be effectively controlled by irrigation and fertilization. Excessive N, which promotes luxurious vegetative growth, increases the risk of blossom-end rot, under the conditions that promote the disorder (drought or a high concentration of soluble salts in the soil).

Magnesium

Pepper needs adequate magnesium (Mg) for normal growth and production of dark green fruit. Mg deficiency is occasionally observed in commercial pepper crops. It most often occurs in plants growing in coarse, sandy soils where Mg fertilizers are not routinely applied or in soils of limestone origin. In severe Mg deficiency, plants fail to grow, and yield is reduced. A marked interveinal chlorosis, which sometimes appears almost white, develops on the older leaves and then the middle-aged leaves (Plate 121). Deficiency symptoms are likely to appear in plants whose leaves have Mg concentrations of less than 0.25%. Where the deficiency is likely to occur, it can be prevented by liming with dolomite or applying Mg-containing fertilizer.

Sulfur

Sulfur (S) deficiency is very rare in pepper crops. Leaves with symptoms of the deficiency are light green to yellow, and the most severe deficiency occurs in recently matured leaves (Plate 122). Symptoms appear in leaves with S concentrations of less than 0.12% (dry whole-leaf basis). Extremely coarse, sandy soils are most often associated with the deficiency. S-containing minerals and fertilizer compounds are common sources of the N, P, and K in fertilizers, and therefore the need for this nutrient is usually satisfied.

Iron

Iron (Fe) deficiency in pepper plants appears in the youngest expanding leaves at the tips of the branches. It causes extreme interveinal yellowing of the leaves. Fe deficiencies are most often observed in crops growing in alkaline soils (pH > 7.0). These soils may occur in areas of limestone geology or where a grower has applied an excessive amount of limestone. Fe deficiency is very difficult to diagnose by leaf analysis, but deficient young leaves usually have Fe concentrations of less than 50 ppm. Foliar Fe sprays are very effective in managing the deficiency.

Manganese

Manganese (Mn) deficiency occurs under conditions similar to those associated with Fe deficiency. Mn deficiency causes interveinal chlorosis or speckling of the young, expanding leaves. In most situations, Mn deficiency is similar to Fe deficiency but not as striking in its symptomatology. Young leaves with Mn concentrations of less than 20 ppm are likely to be deficient in this nutrient.

Selected References

Bosland, P. W., and Votava, E. J. 1999. Peppers: Vegetable and Spice Capsicums. CAB International, Wallingford, U.K.

Kratky, B. A., and Bowen, J. E. 1989. Peppers. Pages 295–303 in: Detecting Mineral Nutrient Deficiencies in Tropical and Temperate Crops. D. L. Pluncknett and H. B. Sprague, eds. Westview Press, Boulder, Colo.

Maynard, D. N., and Hochmuth, G. J. 1997. Knott's Handbook for Vegetable Growers. John Wiley & Sons, New York.

Winsor, G., and Adams, P. 1987. Diagnosis of Mineral Disorders in Plants. Vol. 3, Glasshouse Crops. Her Majesty's Stationery Office, London.

(Prepared by G. J. Hochmuth and D. N. Maynard)

Index

abiotic and physiological disorders, 53–55;
  Pls. 87–103
abnormal fruit shape, 53; Pl. 87
Abutilon mosaic virus, 37
acetolactate synthase inhibitor injury, 57
acifluorfen injury, 58
Acrosternum hilare, 51
Acyrthosiphon
  kondoi, 25
  pisum, 25
alachlor injury, 57
Alfalfa mosaic virus, 24–26, 36; Pl. 48
Alternaria, 43; Pl. 75
  alternata, 43
Alternaria rot, 40, 42–43; Pl. 75
amide herbicide injury, 57
amitrole injury, 58
Andean potato mottle virus, 24
  pepper strain, 26; Pls. 49, 50
anthracnose, 9, 40; Pls. 10–12
aphids
  as pests, 50
  as vectors, 24, 25, 28, 30, 34, 35, 36, 38
Aphis, 36
  craccivora, 28, 35
  fabae, 38
  gossypii, 28, 30, 35, 50
  spiraecola, 28, 35
arthropods, damage caused by, 50–52; Pls.
  82–86
Asparagus virus 2, 25
Athelia rolfsii, 20
atrazine injury, 57; Pl. 107
Aulacorthum solani, 50

Bacillus, 41
  carotovorus, 41
  subtilis, 21
bacteria, diseases caused by, 5–9, 40, 41–42;
  Pls. 1–9, 74
bacterial canker, 5; Pl. 1
bacterial soft rot, 40, 41–42; Pl. 74
bacterial spot, 6; Pls. 2–4
  and sunscald, 55
bacterial wilt, 7–8; Pls. 5–7
banded cucumber beetle, 26
Beet curly top virus, 24, 26–27; Pl. 51
beet leafhopper, 26, 27
Beet western yellows virus, 24
Belladonna mottle virus, 24
Belonolaimus longicaudatus, 47–48
Bemisia
  argentifolia, 52
  tabaci, 28, 29, 31, 32, 37
bensulfuron injury, 57
bensulide injury, 57
bentazon injury, 58; Pl. 114
benzamide herbicide injury, 57
benzoic acid herbicide injury, 57, 58; Pl. 117
bipyridilium herbicide injury, 58; Pl. 113

blossom-end rot, 2, 53; Pls. 88, 89
  and calcium deficiency, 3, 59
  and sunscald, similarity of, 54
blue alfalfa aphid, 25
boron deficiency, 3
Botryotinia fuckeliana, 16
Botrytis cinerea, 11, 16, 17, 43, 44; Pls. 25–
  29, 76, 77
Botrytis fruit rot, 40, 43–44; Pls. 76, 77
  and chilling injury, 45
Broad bean wilt virus, 24, 36
broad mite, 50–51; Pls. 82, 83
bromacil injury, 57
bromoxynil injury, 58
brown stink bug, 51
bud drop, 54
butylate injury, 57

calcium deficiency, 3, 59
  and blossom-end rot, 53
  and Botrytis fruit rot, 44
Capsicum, 1
  annuum, 1
  baccatum, 1
  chinense, 1
  frutescens, 1
  pubescens, 1
Capsicum chlorosis virus, 25, 27
carfentrazone injury, 57, 58; Pl. 112
Carnation Italian ringspot virus, 25
Carrot mottle virus, 24
Celery latent virus, 25
Cercospora capsici, 10; Pl. 13
Cercospora leaf spot. *See* frogeye leaf spot
Chaetanaphothrips signipennis, 51
charcoal rot, 10–11
cheese pepper, 53
Chile blight, 14
Chilli veinal mottle virus, 24, 27–28; Pl. 52
chilling injury, 45
  and Alternaria rot, 40
  and Botrytis fruit rot, 43, 44
Chino del tomate virus, 25, 28–29, 32, 37;
  Pl. 53
chlorsulfuron injury, 57
chlorthal injury, 57
Choanephora blight, 11–12; Pl. 14
Choanephora cucurbitarum, 11; Pl. 14
Circulifer
  opacipennis, 26
  tenellus, 26
Clavibacter michiganensis subsp. *michigan-*
  ensis, 5; Pl. 1
clethodim injury, 57
clomazone injury, 58; Pl. 110
clopyralid injury, 58
Colletotrichum, Pl. 11
  capsici, 9, 40
  coccodes, 9
  gloeosporioides, 9, 40

color spotting (abiotic disorder), 53–54;
  Pl. 90
contact action herbicide injury, 57, 58; Pls.
  111–114
cookie fruit, 53
Corticium rolfsii, 20
Corynebacterium michiganense, 5
cotton aphid, 50
cracking, of fruit, 54
Criconemella, 48
Cucumber mosaic virus, 24, 29–31, 36; Pls.
  54–56
cultural practices, in pepper production, 2–3
Cuscuta, 45; Pl. 78
Cymbidium ringspot virus, 25
Cytophaga, 41

2,4-D injury, 57, 58
2,4-DB injury, 58
damping-off, 2, 12–13; Pl. 15
DCPA injury, 57
Diabrotica balteata, 26
dicamba injury, 57, 58; Pl. 117
diclosulam injury, 57
dinitroaniline herbicide injury, 57
diphenamid injury, 57
diphenyl ether herbicide injury, 58; Pl. 111
diquat injury, 58
diuron injury, 57
dodder, 45–46; Pl. 78
Dolichodorus heterocephalus, 48

Eggplant mosaic virus, 24
Eggplant mottled crinkle virus, 25
Eggplant mottled dwarf virus, 25
Eggplant severe mottle virus, 24
EPTC injury, 57
Erwinia
  aroideae, 41
  carotovora subsp. *atroseptica,* 41
  carotovora subsp. *carotovora,* 41
  chrysanthemi, 41
ethephon, 3
Euschistus servus, 51

false root-knot nematode, 48
fertilization, 2–3, 59, 60
fish-mouth (seed disorder), 4
fleahoppers, 51, 52
flower and flower bud drop, 3, 54; Pl. 91
fluazifop-P injury, 57
fomesafen injury, 58
foxglove aphid, 50
Frankliniella
  bispinosa, 40
  fusca, 40
  intonsa, 40
  occidentalis, 27, 40, 51
  schultzei, 27, 40
frogeye leaf spot, 10; Pl. 13

fruit cracking, 54
fruit drop, 3, 54
fruit malformation (abiotic disorder), 53
fungi and oomycetes, diseases caused by, 9–23, 40–41, 42–45; Pls. 10–47, 75–77
Fusarium, 12
  annuum, 14
  oxysporum, 14; Pls. 18–20
  oxysporum f. sp. *vasinfectum,* 14
  solani, 13; Pls. 16, 17
  vasinfectum, 14
Fusarium stem rot, 13; Pls. 16, 17
Fusarium wilt, 14–15; Pls. 18–20

Geotrichum candidum, 40
gibberellic acid, 3
glasshouse potato aphid, 50
Gliocladium virens, 21
Glomerella cingulata, 9
glufosinate injury, 57, 58
glyphosate injury, 57, 58; Pls. 115, 116
gray leaf spot, 15–16; Pls. 21–24
gray mold, 16–17; Pls. 25–29
gray mold rot. *See* Botrytis fruit rot
green peach aphid. *See Myzus persicae*
green stink bug, 51
greenhouse production, 3
greenhouse whitefly, 52
growth regulators, 3

hail injury, 54; Pls. 92, 93
halosulfuron injury, 57; Pl. 104
Halticus, 51
Helicotylenchus dihystera, 48
Hemicycliophora arenaria, 48
Henbane mosaic virus, 24
herbicide injury, 56–58; Pls. 104–118
hexazinone injury, 57; Pl. 108
hydroponic culture, 3
Hysteroneura setariae, 28

imazethapyr injury, 57; Pl. 105
imidazolinone herbicide injury, 57; Pl. 105
Impatiens necrotic spot virus, 25
Indian peanut clump virus, 24
Indian pepper mottle virus, 24
insects and mites, damage caused by, 50–52; Pls. 82–86
iron deficiency, 60
irrigation, 2
isoxaben injury, 57

lactofen injury, 58
leafhoppers, as vectors, 26, 27
lesion nematode, 48
Leveillula taurica, 19
linuron injury, 57
Lychnis ringspot virus, 25
Lygus lineolaris, 51

Macrophomina phaseolina, 10
Macrosiphum euphorbiae, 38, 50
magnesium deficiency, 3, 59; Pl. 121
malformation of fruit (abiotic disorder), 53
manganese deficiency, 60
manganese toxicity, 3
marigold mottle virus, 24
MCPA injury, 58
Meloidogyne, Pls. 79–81
  arenaria, 46, 47
  hapla, 46, 47
  incognita, 46, 47; Pl. 79
  javanica, 46, 47
melon aphid, 50
melon thrips, 27

metolachlor injury, 57
metribuzin injury, 57
metsulfuron injury, 57
molinate injury, 57
Moroccan pepper virus, 25
MSMA injury, 58
mulches, plastic, 2
Myzus persicae, 25, 28, 34, 35, 36, 38, 50

Nacobbus aberrans, 48
napropamide injury, 57
Nectria haematococca, 13
nematodes, diseases caused by, 46–49; Pls. 79–81
Nezara viridula, 51
nicosulfuron injury, 57
nitrogen, as nutrient, 3, 59
norflurazon injury, 58; Pl. 109
nutritional disorders, 3, 59–60; Pls. 119–122

Oidiopsis
  sicula, 19; Pls. 35–38
  taurica, 19
onion thrips, 51
Ourmia melon virus, 25
oxyfluorfen injury, 57, 58; Pl. 111

packaging, of pepper fruit, 42, 45
Paprika mild mottle virus, 25
paraquat injury, 57, 58; Pl. 113
Paratrichodorus, 48
  minor, 48, 49
pea aphid, 25
Pea early-browning virus, 24
Peanut stunt virus, 24
Pectobacterium carotovorum, 41
Pelargonium vein clearing virus, 25
Pellicularia rolfsii, 20
pendimethalin injury, 57
Penicillium, 21
pepper, culture of, 1–4
Pepper golden mosaic virus, 25, 31, 32; Pls. 57, 58
Pepper huasteco virus, 25, 32
Pepper huasteco yellow vein virus, 25, 32
Pepper mild mosaic virus, 24
Pepper mild mottle virus, 25, 32–33, 39; Pl. 59
Pepper mild tigre virus, 25, 31; Pl. 58
Pepper mottle virus, 24, 26, 33–34, 35, 38; Pl. 60
Pepper ringspot virus, 24
Pepper severe mosaic virus, 24
Pepper vein banding virus, 24
Pepper veinal mottle virus, 24, 28, 34–35; Pl. 61
Peru tomato mosaic virus, 24
Petunia asteroid mosaic virus, 25
phenoxy acid herbicide injury, 58
Physalis mosaic virus, 25
Physalis mottle virus, 24
physiological disorders, 53–55; Pls. 87–91
Phytophthora, 12
  capsici, 11, 14, 17, 18, 19, 40; Pls. 30–34
  nicotianae var. *parasitica,* 41
Phytophthora blight, 17–19, 40; Pls. 30–34
  and bacterial wilt, similarity of, 7
  and sunscald, 55
picloram injury, 57, 58
Pittosporum vein yellowing virus, 25
plant bugs, 51
Polyphagotarsonemus latus, 50; Pls. 82, 83
postharvest diseases and disorders, 40–45; Pls. 74–77

potassium deficiency, 3, 59; Pls. 119, 120
potato aphid, 50
Potato aucuba mosaic virus, 24
Potato leaf roll virus, 24
Potato mop-top virus, 24
Potato virus A, 24
Potato virus X, 25
Potato virus Y, 24, 33, 35–36, 38; Pls. 62–64
powdery mildew, 19–20; Pls. 35–38
  and sunscald, 55
Pratylenchus, 48
  penetrans, 48
propanil injury, 58
Pseudomonas, 12, 41
  marginalis, 41
  solanacearum, 7
  syringae pv. *syringae,* 8; Pls. 8, 9
pyridine herbicide injury, 58
pyrithiobac injury, 57; Pl. 106
Pythium
  aphanidermatum, 12
  myriotylum, 12
  ultimum, 12

quinclorac injury, 57, 58; Pl. 118
quinolinecarboxylic acid herbicide injury, 58; Pl. 118

Radopholus similis, 48
raised beds, for pepper production, 2
Ralstonia
  eutropha, 7
  pickettii, 7
  solanacearum, 7; Pls. 5–7
reniform nematode, 48
Rhizoctonia solani, 12
Rhizopus
  nigricans, 44
  stolonifer, 44, 45
Rhizopus rot, 40, 41, 44–45
Rhopalosiphum maidis, 28
Ribgrass mosaic virus, 25
root-knot nematodes, 46–47; Pls. 79–81
  and sunscald, 55
Rotylenchulus reniformis, 48

salt injury, 54; Pls. 94–96
  and damping-off, similarity of, 12
Scirtothrips dorsalis, 51
Sclerotinia rot. *See* white mold
Sclerotinia sclerotiorum, 23; Pls. 45–47
Sclerotium rolfsii, 20; Pls. 39, 40
seed disorders, 3–4, 12
seed production, 3
seed treatments, 4
Serrano golden mosaic virus, 25, 31, 32; Pl. 57
sethoxydim injury, 57
silverleaf whitefly, 52
simazine injury, 57
Sinaloa tomato leaf curl virus, 25, 32, 36–37; Pl. 65
sodium, sensitivity of peppers to, 3. *See also* salt injury
soil, for pepper production, 2
sour rot, 40
southern blight, 20–21; Pls. 39, 40
southern green stink bug, 51
spotting (abiotic disorder), 53
Stemphylium, Pls. 21–24
  botryosum f. *capsici,* 15
  floridanum, 15
  lycopersici, 15
  solani, 15
sting nematode, 47–48

stink bugs, 51, 52; Pl. 86
storage, of pepper fruit, 42, 43, 45
stubby-root nematodes, 48
substituted urea herbicide injury, 57
sulfentrazone injury, 58
sulfonylurea herbicide injury, 57; Pl. 104
sulfur deficiency, 60; Pl. 122
Sunn-hemp mosaic virus, 25
sunscald, 54–55; Pls. 97–99
   and bacterial soft rot, 41
syringae seedling blight and leaf spot, 8–9;
    Pls. 8, 9

tarnished plant bug, 51, 52
temperature, in storage, 45
terbacil injury, 57
Texas pepper virus, 25, 31, 32; Pls. 57, 58
thiocarbamate herbicide injury, 57
Thrips
   hawaiiensis, 51
   palmi, 27, 51
   parvispinus, 51
   setosus, 40
   tabaci, 40, 51
thrips
   as pests, 51; Pls. 84, 85
   as vectors, 25, 27, 40
tillage, in pepper production, 2
Tobacco etch virus, 24, 26, 36, 38; Pls. 66, 67

Tobacco leaf curl virus, 25
Tobacco mild green mosaic virus, 25
Tobacco mosaic satellite virus, 25
Tobacco mosaic virus, 25, 33, 38–39
Tobacco necrosis virus, 24
Tobacco rattle virus, 24
Tobacco ringspot virus, 25
Tobacco streak virus, 25
tobacco thrips, 40
Tomato aspermy virus, 24
Tomato bushy stunt virus, 25
Tomato leaf crumple virus, 28, 29
Tomato mosaic virus, 25, 33, 38–39; Pls.
   68, 69
Tomato mottle virus, 37
Tomato necrotic dwarf virus, 25
Tomato ringspot virus, 25
Tomato spotted wilt virus, 25, 27, 39–40; Pls.
   70–73
tomato thrips. See *Frankliniella schultzei*
Toxoptera citricida, 28
transplanting, 2
Trialeurodes vaporariorum, 52
triazine herbicide injury, 57; Pl. 107
triazinone herbicide injury, 57, 58; Pls. 108,
   112
Trichoderma
   harzianum, 20, 21
   viride, 21

Trichodorus, 48
triclopyr injury, 58
trifluralin injury, 57
true bugs, 51–52; Pl. 86

uracil herbicide injury, 57

vernolate injury, 57
Verticillium, Pls. 41–44
   albo-atrum, 21, 22
   dahliae, 21, 22
Verticillium wilt, 21–22; Pls. 41–44
   and sunscald, 55
viruses, diseases caused by, 23–40; Pls. 48–73

western flower thrips, 27, 40, 51
wet rot, 11–12; Pl. 14
white mold, 22–23; Pls. 45–47
whiteflies
   and broad mite infestations, 50, 51
   as pests, 52
   as vectors, 28, 29, 31, 32, 37
wind injury, 55; Pls. 100–103

Xanthomonas, Pls. 2–4
   campestris, 41
   campestris pv. *vesicatoria,* 6
   vesicatoria, 6